Controlled Drug Analysis

Controlled Drug Analysis

By

Michael D. Cole
Anglia Ruskin University, UK
Email: michael.cole@aru.ac.uk

Lata Gautam
Anglia Ruskin University, UK
Email: lata.gautam@aru.ac.uk

and

Agatha Grela
Anglia Ruskin University, UK
Email: agatha.grela1@aru.ac.uk

Print ISBN: 978-1-78801-534-9
PDF ISBN: 978-1-83767-305-6
EPUB ISBN: 978-1-78801-935-4

A catalogue record for this book is available from the British Library

© Michael D. Cole, Lata Gautam and Agatha Grela 2025

All rights reserved

Apart from fair dealing for the purposes of research for non-commercial purposes or for private study, criticism or review, as permitted under the Copyright, Designs and Patents Act 1988 and the Copyright and Related Rights Regulations 2003, this publication may not be reproduced, stored or transmitted, in any form or by any means, without the prior permission in writing of The Royal Society of Chemistry or the copyright owner, or in the case of reproduction in accordance with the terms of licences issued by the Copyright Licensing Agency in the UK, or in accordance with the terms of the licences issued by the appropriate Reproduction Rights Organization outside the UK. Enquiries concerning reproduction outside the terms stated here should be sent to The Royal Society of Chemistry at the address printed on this page.

Whilst this material has been produced with all due care, The Royal Society of Chemistry cannot be held responsible or liable for its accuracy and completeness, nor for any consequences arising from any errors or the use of the information contained in this publication. The publication of advertisements does not constitute any endorsement by The Royal Society of Chemistry or Authors of any products advertised. The views and opinions advanced by contributors do not necessarily reflect those of The Royal Society of Chemistry which shall not be liable for any resulting loss or damage arising as a result of reliance upon this material.

The Royal Society of Chemistry is a charity, registered in England and Wales, Number 207890, and a company incorporated in England by Royal Charter (Registered No. RC000524), registered office: Burlington House, Piccadilly, London W1J 0BA, UK, Telephone: +44 (0) 20 7437 8656.

For further information see our website at www.rsc.org

For general enquiries, please contact books@rsc.org

For EU product safety enquiries, please email books@rsc.org
or contact Royal Society of Chemistry Worldwide (Germany) GmbH,
Römischer Hof, Unter den Linden 10, 10117 Berlin.

Printed in the United Kingdom by CPI Group (UK) Ltd, Croydon, CR0 4YY, UK

Dedication

Michael D. Cole dedicates this book to his mother, who saw the book started, but sadly didn't see it complete. Her love for books was an inspiration.

Lata Gautam dedicates this book to her grandmother who passed away after having a successful and happy 97 years.

Agatha Grela dedicates this book to Clara and Joshua Jun, who are the future.

Preface

The use, misuse and abuse of recreational drugs is a contentious issue, with some arguing that drug control should remain in place or become even tighter, while others argue for a relaxation or the abolition of drug control legislation altogether. It is not the intention of this book to present arguments for and against each of these positions. Regardless of the side of the argument that someone takes, while legislation is in place, certain drugs will be controlled by the legal process and court cases will follow. Consequently, forensic scientists will be presented with controlled substances around which a number of questions may be asked by those involved in the legal process. Even a seemingly simple question such as "Is there a controlled substance present?" may be easy or difficult to answer, depending upon what the material is. Some drugs are extremely well-worked, for example, *Cannabis* products, heroin and cocaine. The dynamic and international nature of the recreational drug market, coupled with widespread access to the internet, does mean, however, that new classes of drugs and mixtures of drugs are constantly emerging. Examples include synthetic cannabinoid receptor agonists (SCRAs), for many of which standard reference materials were not available when the drugs first were available in the market. More recently, nitazenes are exercising the minds of forensic drug chemists. Further, polydrug mixtures are emerging, which renders their analysis difficult. One such example is Nyaope, a South African drug containing cannabis, heroin and antiretroviral drugs. The techniques for one class of drugs in the mixture may not be appropriate for the analysis of another class of drugs present within the same mixture. Added to this, the drug mixtures may change with time, in the case of Nyaope, through the addition of tramadol by those who prepare the drug mixtures before sale. More complicated questions may then arise – for example, How much drug is present? How was the drug made? How do the samples in a seizure or group of seizures relate to each other?

Controlled Drug Analysis
By Michael D. Cole, Lata Gautam and Agatha Grela
© Michael D. Cole, Lata Gautam and Agatha Grela 2025
Published by the Royal Society of Chemistry, www.rsc.org

There are often multiple ways to address these questions in terms of the analysis of the drugs. These multiple ways are required because not every laboratory in which the analysis of drugs is carried out has the same level of resources, equipment, personnel or even electricity for 24 hours a day. This means that different analytical methods have to be used in different laboratories or the materials be sent away for analysis. The latter approach suffers from the disadvantages that the materials may be lost, sample integrity may be compromised, or the materials may change chemically in some way through hydrolysis, oxidation, decomposition in light or artefact formation.

On this basis, it is important for those involved in drug seizure, analysis and evidence presentation to be aware of the different approaches that might be adopted to address the questions that may be posed around a drug sample. This book is not a "recipe book". Numerous analytical methods are presented in other texts and extremely comprehensively in the "Recommended Methods" series published by the United Nations Office on Drugs and Crime. This book aims, through a discussion of a number of different types of drugs and drug classes, to address the principles and philosophy of drug analysis, to help the reader address the following questions:

- Have I got a sample from which I can obtain an answer to the question that I am being asked?
- Is the method of sampling and analysis that I am using appropriate?
- Are there alternative methods that I can use, and if so, would they be better than my chosen method?
- Have the necessary controls been analysed?

These are addressed throughout the book.

The Sydney Declaration[1] is particularly pertinent to drug analysis and presentation of relevant testimony in court. The Declaration states that Forensic Science is

"A case based, research orientated science based endeavour to study traces – the remnants of past activities – through the detection, recognition, recovery, examination and interpretation of anomalous events of public interest"

In the context of the Sydney Declaration, drug analysis is driven by casework, often requiring research on new and unknown drugs, where the components of drug mixtures must be recovered, detected, recognised and examined (identified) and the data correctly interpreted and reported through a due legal process. It is this, and the dynamic nature of the drugs market, that makes drug analysis so interesting and intellectually challenging. Additionally, a wide range of researchers are making contributions to the knowledge base around drug analysis. Their material can be found in outputs relating to diverse disciplines, including chemistry, forensic science, biochemistry, botany, phytochemistry, law, education and management.

Preface

While this list is not exhaustive, it illustrates the diversity of disciplines that contribute to drug analysis, which only adds to the fascinating nature of the subject.

It is hoped that the information contained within this book would allow a forensic scientist and others involved in law enforcement and the legal process to be able to (i) carry out drug analyses appropriately and (ii) critically appraise any drug-related evidence that might be presented to court. Additionally, should a new drug class be encountered (which will inevitably happen), they should be able to (i) design an analytical sequence appropriate to the chemistry of the drug and the resources available to the lab and/or (ii) evaluate the evidence so generated. The drug classes encountered will change with the drug market, but the principles and philosophy behind the analysis should not, remembering always that, ultimately, the role of the forensic scientist is to provide accurate information to a court to allow it to make an informed decision.

Michael D. Cole
Lata Gautam
Agatha Grela
Cambridge, England

Reference

1. C. Roux, R. Bucht, F. Crispino, P. De Forest, C. Lennard, P. Margot, M. D. Miranda, N. NicDaeid, O. Ribaux, A. Ross and S. Willis, *Forensic Sci. Int.*, 2022, **332**, 111182.

Acknowledgements

This book has been in gestation and writing for a long time. It brings together, in our collective experience, just under 60 years of teaching, casework and research in forensic science in general, and drug chemistry in particular, which has been undertaken in more than 40 countries worldwide. Over these journeys, there have been innumerable conversations, and each has influenced, in some large way or small, our approach to controlled drug analysis. We must acknowledge and thank all of those who have engaged with us in these conversations that have influenced how we think today. We are fortunate enough to work with university students in both taught and research programmes. Our students are not afraid to ask us those probing questions that cause you to stop and think about how you do something or, often and more importantly, why. Again, our thanks are due to our students, who we hope will continue to ask those challenging questions.

This book would not have been possible without the sustained support of a number of people. We thank Janet Freshwater and Liangqi Li and colleagues at the Royal Society of Chemistry for their patient support to get this book together. We thank Professor Chris Parris, Head of Life Sciences at ARU, for his continued support in this endeavour, as well as our colleagues at ARU. We also thank Jana Kollarova for assistance in drawing chemical structures and for the critical, thought-provoking conversations about how drug chemistry, and more broadly forensic chemistry, can be taught to different learner types.

Finally, we thank our respective friends and families for their support, encouragement, and patience while we locked ourselves away to write.

Contents

1	**Legislative Systems and Controlled Drugs**	**1**
	1.1 Introduction	1
	1.2 Drug Naming	2
	1.3 International Drug Control	3
	1.4 Drug Control in the United Kingdom	3
	1.5 Drug Control in the United States of America	6
	1.6 Drug Control in Australia	8
	1.7 Drug Control in India	8
	1.8 Conclusion	8
	References	9
2	**Drug Sampling**	**10**
	2.1 Introduction	10
	2.2 Physical Description	11
	2.3 Trace or Bulk: What Should Be Done?	13
	2.3.1 Recovering Drugs from Trace Samples	13
	2.3.2 Treatment of Bulk Samples	16
	2.3.3 Other Considerations	17
	2.4 Packaging of Drug Materials	18
	2.5 The Number of Samples to Take	20
	2.5.1 Arbitrary Sampling	21
	2.5.2 Statistical Sampling	21
	2.6 Representative Sampling of Drugs	22
	2.6.1 Sampling of Powders	22
	2.6.2 Sampling of Liquids	23

Controlled Drug Analysis
By Michael D. Cole, Lata Gautam and Agatha Grela
© Michael D. Cole, Lata Gautam and Agatha Grela 2025
Published by the Royal Society of Chemistry, www.rsc.org

	2.6.3 Sampling of Tabletted Dose Forms	24
	2.6.4 Sample Dissolution	25
	References	25

3 The Analysis of Cannabis and Products — 27

3.1 Introduction — 27
    3.1.1 Cannabis — 27
3.2 Analytical Sequence for the Analysis of Cannabis — 29
    3.2.1 Bulk *vs.* Trace Samples in the Analysis of Cannabis Samples — 30
    3.2.2 Physical Description of Cannabis-containing Materials and Associated Paraphernalia — 30
    3.2.3 Sampling Cannabis Materials — 35
    3.2.4 Chemical Analysis of Cannabis — 36
References — 47

4 The Analysis of Synthetic Cannabinoids — 48

4.1 Introduction — 48
4.2 Analysis of Synthetic Cannabinoids — 53
    4.2.1 Physical Description — 53
    4.2.2 Presumptive Tests for Synthetic Cannabinoids — 53
    4.2.3 Thin Layer Chromatography of Synthetic Cannabinoids — 54
    4.2.4 Gas Chromatography of Synthetic Cannabinoids — 56
    4.2.5 High-performance Liquid Chromatographic Methods — 58
4.3 NMR Analysis of Synthetic Cannabinoids — 60
References — 61

5 The Analysis of Hallucinogenic Drugs from Plants and Fungi — 62

5.1 Introduction — 62
5.2 The Analysis of LSD — 63
    5.2.1 Physical Description of LSD — 66
    5.2.2 Sampling of Dose Units Expected to Contain LSD — 66
    5.2.3 Presumptive Tests for LSD — 67
    5.2.4 Thin Layer Chromatography of LSD and Related Drugs — 67
    5.2.5 High-performance Liquid Chromatography of LSD — 68
    5.2.6 Gas Chromatography–Mass Spectrometry of LSD — 69
    5.2.7 Spectroscopic Methods for the Analysis of LSD — 70

5.3 Analysis of Fungal Material and Drugs Containing Psilocybin and Psilocin	72
5.3.1 Packaging, Handling and Physical Examination of Fungal Samples	74
5.3.2 Physical Examination of Drugs of Fungal Origin	74
5.3.3 Extraction and Presumptive Testing of Mushroom-based Drugs	75
5.3.4 Thin Layer Chromatography of Drugs of Fungal Origin	76
5.3.5 Gas Chromatography–Mass Spectrometry	76
5.3.6 High-performance Liquid Chromatography	77
5.3.7 DNA Analysis of Fungal Material	78
5.4 The Analysis of DMT	79
5.4.1 Introduction	79
5.4.2 Presumptive Tests for DMT	80
5.4.3 Thin Layer Chromatography of DMT	80
5.4.4 Gas Chromatography–Mass Spectrometry of DMT	80
5.4.5 High-performance Liquid Chromatography of DMT	81
5.4.6 Nuclear Magnetic Resonance Spectroscopy of DMT	82
5.5 Analysis of Mescaline	82
5.5.1 Packaging, Storage and Sampling of Cactus Material	83
5.5.2 Extraction of Mescaline	83
5.5.3 Presumptive Tests for Mescaline	84
5.5.4 Thin Layer Chromatography of Mescaline	84
5.5.5 Gas Chromatography of Mescaline	85
References	86

6 The Analysis of Khat and Cathinones   87

6.1 Introduction	87
6.2 Forms, Dosage and Effect of Synthetic Cathinones	92
6.3 Identification of Khat and Cathinones	93
6.3.1 Identification of Plant Material as Khat (*Catha edulis*)	93
6.3.2 Identification of Synthetic Cathinones	94
6.4 Other Analytical Techniques	100
6.5 Profiling of Cathinones	101
References	102

7 The Analysis of Opiate Drugs and Heroin   103

7.1 Introduction	103
7.2 The Origins of Heroin	106
7.3 The Manufacture of Heroin	106

7.4 Physical Description of Heroin Samples	108
7.5 Presumptive Tests for Heroin Samples	109
7.6 Thin Layer Chromatography of Heroin	109
7.7 High-performance Liquid Chromatography of Heroin	111
7.8 Gas Chromatography–Mass Spectrometry of Heroin	113
7.9 NMR Spectroscopy of Heroin Samples	116
7.10 The Analysis of Stable Isotopes in Heroin	118
7.11 The Analysis of Metal Ions in Heroin	120
7.12 The Analysis of Occluded Solvents in Heroin	120
7.13 DNA Analysis and Heroin	123
References	123

8 The Analysis of Amphetamines, Ring-substituted Amphetamines and Related Compounds — 125

8.1 Introduction	125
8.2 Synthetic Routes for Amphetamine Synthesis	127
8.3 Description and Sampling of Amphetamine-type Drugs	131
8.4 Presumptive Tests for Amphetamine-type Drugs	132
8.5 Thin Layer Chromatography of Amphetamine-type Drugs	133
8.6 High-performance Liquid Chromatography of Amphetamine-type Drugs	135
8.7 Gas Chromatography of Amphetamines	138
8.7.1 Qualitative Analysis of Underivatised Amphetamines	139
8.7.2 Identification of Amphetamines by GC–MS	140
8.8 The Analysis of Inorganic Ions in Samples of Amphetamine and Related Drugs	142
8.9 Chemical Impurity Profiling of Amphetamine-type Drugs	143
References	146

9 The Analysis of Barbiturate Drugs — 148

9.1 Introduction	148
9.2 Presumptive Tests for Barbiturate Drugs	150
9.3 Thin Layer Chromatography of Barbiturate Drugs	151
9.4 High-performance Liquid Chromatography of Barbiturates	151
9.5 Gas Chromatographic Analysis of Barbiturates	152
9.6 NMR Spectroscopy of Barbiturates	154
References	154

Contents xvii

10 The Analysis of Phenyl- and Benzylpiperazines   155

10.1 Introduction   155
10.2 Synthesis of Piperazines   158
10.3 The Forensic Examination of Piperazines   159
    10.3.1 Physical Description of Piperazine-based Drugs   159
    10.3.2 Screening Tests for Piperazines   159
    10.3.3 Dissolving Piperazines for Analysis   160
    10.3.4 Thin layer Chromatography of Piperazines   162
    10.3.5 Instrumental Methods for the Analysis of Piperazines   163
References   166

11 The Analysis of Cocaine   168

11.1 Introduction   168
11.2 The Extraction and Synthesis of Cocaine from Plant Material   170
11.3 Analysis of Cocaine   171
    11.3.1 Description and Sampling of Cocaine   171
    11.3.2 Presumptive Tests for Cocaine   172
    11.3.3 Microcrystalline Tests for Cocaine   172
    11.3.4 Preparation of Solutions for the Analysis of Cocaine   173
    11.3.5 Thin Layer Chromatography of Cocaine   174
    11.3.6 Gas Chromatography–Mass Spectrometry of Cocaine   175
    11.3.7 High-performance Liquid Chromatography of Cocaine   176
    11.3.8 Ultraviolet Spectroscopy of Cocaine   177
    11.3.9 Infrared Spectroscopy of Cocaine   178
    11.3.10 Analysis of DNA in Cocaine   179
    11.3.11 Stable Isotope Analysis and Cocaine   179
References   180

12 The Analysis of Benzodiazepines   181

12.1 Introduction   181
12.2 Analysis of Benzodiazepines   186
    12.2.1 Sampling and Physical Description of Benzodiazepines   186
    12.2.2 Presumptive Tests for Benzodiazepines   187
    12.2.3 Dissolution of Benzodiazepines for Further Analysis   187
    12.2.4 Thin Layer Chromatography of Benzodiazepines   188
    12.2.5 Gas Chromatographic-based Analysis of Benzodiazepines   189

	12.2.6 High-performance Liquid Chromatography-based Techniques	191
References		192

13 The Analysis of Fentanyl and Its Analogues — 193

13.1 Introduction 193
13.2 Analytical Methods for the Analysis of Fentanyl-type Drugs 196
    13.2.1 Sampling of Fentanyl-type Drugs 196
    13.2.2 Presumptive Testing of Fentanyl-type Drugs 197
    13.2.3 Instrumental Methods for Screening for Fentanyl-type Drugs 198
    13.2.4 Confirmation of Identity and Quantification of Fentanyl-type Drugs 199
References 201

14 The Analysis of Phencyclidine and Ketamine — 202

14.1 Introduction 202
14.2 Analysis of Phencyclidine and Ketamine 204
    14.2.1 Sampling and Presumptive Tests for Ketamine and PCP 204
    14.2.2 Presumptive Tests for Phencyclidine and Ketamine 204
    14.2.3 Microcrystal Tests for Phencyclidine and Ketamine 205
    14.2.4 Thin Layer Chromatography of PCP and Ketamine 205
    14.2.5 Instrumental Analysis of Phencyclidine and Ketamine 206
    14.2.6 Identification of New Analogues of Ketamine 208
References 208

Subject Index — 209

1 Legislative Systems and Controlled Drugs

1.1 Introduction

The use and abuse of recreational drugs continue to increase, with new types of drugs and mixtures of drugs constantly emerging. In the UK, for example, at the time of writing, nitazenes are proving to be problematic. In other parts of the world, new polydrug mixtures – for example, Happy Water, Fairy Water, Nyaope and White Pipe – are proving to be analytically challenging. Regardless of personal views, and the fact that drugs such as cannabis, cocaine and opium-derived products have been used for thousands of years, forensic scientists work within legislative frameworks and provide the required information necessary for legal processes to be implemented. It is therefore necessary for those involved in drug analysis to at least have a working knowledge of the legislative system in which they are working. The problems are compounded by the fact that new drugs are constantly emerging, new combinations of drugs are being encountered, and legislation surrounding controlled substances is changing, with increasing regulation in some cases and deregulation in others. For example, Uruguay, Canada and a number of US states have accepted to legalise recreational cannabis. Furthermore, some legislation, for example the Misuse of Drugs Act (MDAct) in the United Kingdom, is more than 50 years old and has required amendment as drug use has changed. There is also a responsibility on the forensic scientist to fully understand the analytical methods and to fully and critically evaluate the analytical methodology used, and data obtained. Some countries have the death penalty for drugs related offences, including China, Dubai, Indonesia,

---

Controlled Drug Analysis
By Michael D. Cole, Lata Gautam and Agatha Grela
© Michael D. Cole, Lata Gautam and Agatha Grela 2025
Published by the Royal Society of Chemistry, www.rsc.org

Iran, Malaysia, the Philippines, Saudi Arabia, Singapore, Thailand and Vietnam, while others legislate for extremely long prison sentences. Of course, regardless of the legislative infrastructure, the duty of a forensic scientist is to help a court arrive at an informed decision and so it is essential that findings are sound and grounded in good science. The following discussion provides the basis for understanding a number of the different types of legislative systems that are in operation and a framework for the analytical methods that are described in later chapters in this book.

1.2 Drug Naming

It is essential for forensic drug chemists to understand the drug naming system within the legislative framework in which they are working. Some legislation uses IUPAC names, some legislative instruments use standard non-proprietary names, and some use colloquial names. The use of colloquial names is problematic. For example, the use of the name Kush in the United States refers to synthetic cannabinoid receptor agonists dried onto tobacco. However, in West Africa (for example, Sierra Leone, Liberia, and Guinea), Kush refers to a mixture of tobacco, cannabis, fentanyl and tramadol.[1]

In some systems, common names are used interchangeably with non-proprietary names. For example, diamorphine and heroin are used interchangeably in a number of legislative instruments. In other systems, heroin is taken to mean the mixture of compounds resulting from the synthesis of diamorphine from morphine with anything that has been added to the drug, while diamorphine is used to refer to the single chemical. To complicate matters further, some systems consider the whole drug seizure as a drug. Again, using heroin as the example, a seizure of 1 g of diamorphine in 99 g of non-drug material would be considered, in some systems, to be equivalent to 100 g of diamorphine.

The use of street names may be problematic because the composition of the drug may change over time. For example, Nyaope, also known as Whoosh and Unga, a drug mixture found in South Africa, originally contained mixtures of tobacco, cannabis, heroin and anti-HIV antiretrovirals, some of which are known to be hallucinogenic. Recent seizures (2022–2024) have also been found to contain hydromorphone and tramadol, both of which were not recorded in earlier samples.[2] Given the problems associated with the names of drugs, it is essential for drug chemists to be absolutely clear about which drug or drugs they are referring to and the output of their analyses.

1.3 International Drug Control

Modern international and national legislation around controlled substances is based upon three United Nations treaties: the Single Convention on Narcotic Drugs 1961, the Convention on Psychotropic Substances 1971, and the Convention against Illicit Traffic in Narcotic Drugs and Psychotropic Substances 1988. Signatories to these treaties implement them through domestic laws and legislative instruments. In essence, these instruments control the possession, supply, trafficking and production of drugs generally listed in schedules according to their harmfulness. Interestingly, "psychotropic" is neither defined in the 1971 instrument nor in many of the national systems. The 1988 instrument is also important in tackling organised crime and allows the seizing of drug-related assets.

1.4 Drug Control in the United Kingdom

The 1971 UN instrument was brought into UK law by the Misuse of Drugs Act 1971 (MDAct). In this instrument, controlled drugs are set out in Schedule 2 of the MDAct. Activities including possession, supply, production, importation and export, manufacture and allowing the use of drug taking are prohibited under different sections of the Act, including

> Section 3: Importation and exportation of controlled substances,
> Section 4: Production, supply or offer to supply a controlled substance,
> Section 5: Possession of a controlled substance,
> Section 6: Cultivation of any plant of the genus *Cannabis*,
> Section 8: Knowingly permitting activities listed in the Act to occur as an occupier or manager of premises on which the activities occur, and

Section 9: Preparation and smoking of opium, the frequenting of a place where this occurs and being in possession of pipes or utensils adapted for the smoking of opium.

Within the MDAct, drugs are listed either by name or generically by chemical group or type and are divided into Class A, B and C drugs according to their harmfulness and the risks associated with them. In addition to named drugs and drug groups, stereo and positional isomers are generically controlled as are different salt forms. The full wording of the current version of the Misuse of Drugs Act can be found on the UK Government website.[3] The penalties for offences relating to drugs listed in the MDAct are given in Table 1.1. Legislation in this form means that the forensic scientist must provide, firstly, definitive identification of any controlled substance(s) present in a sample. Secondly, in order to provide an opinion on whether a case may be one of possession or of supply, the amount of drug present must be determined; it is also necessary to have a working knowledge of the ranges of the amount of drug taken by users, from a recreational occasional user to a heavy user of the drug. From such data calculations can be carried out to determine whether the amount of drug present might be for personal use or for supply. Whether or not the drug decomposes over time is also important in determining the possible explanation for the possession of different amounts of drugs. Sometimes, the drug chemist is also asked about the value of the drugs at the street level, and so it is necessary to have a working knowledge of the sale price of drugs in different regions. Additionally, they must have a working knowledge of drug synthesis and manufacture. For example, they may be asked whether or not a drug might be synthesised from chemicals that have been seized. They may also be asked whether impurities in drugs may be

Table 1.1 Maximum penalties upon conviction under the Misuse of Drugs Act 1971.

Class	Possession	Supply/production
A	Up to 7 years in prison, an unlimited fine or both	Up to life in prison, an unlimited fine or both
B	Up to 5 years in prison, an unlimited fine or both	Up to 14 years in prison, an unlimited fine or both
C	Up to 2 years in prison, an unlimited fine or both	Up to 14 years in prison, an unlimited fine or both
Temporary class	Police can confiscate temporary class drugs	Up to 14 years in prison, an unlimited fine or both

used to relate drug samples to each other in order to address supply and trafficking charges.

In addition to the MDAct 1971, the Misuse of Drugs Act (Regulations) 1985 and 2001 provide further legislative instruments that drugs chemists should be aware of. The 1985 Regulations, for the first time, control drugs such as benzodiazepines. The Regulations also describe what is allowed under different licensing conditions. The Drug Trafficking Offences Act 1986 makes provision for the confiscation of proceeds from the trafficking of controlled substances. In addition, over the years, a large number of novel psychoactive substances have appeared in the illicit drug market. The United Nations Office on Drugs and Crime has defined psychoactive substances as:

> substances of abuse, either in pure form or in a preparation, that are not controlled by the 1961 Single Convention on Narcotic Drugs or the 1971 Convention on Psychotropic Substances but which may pose a public health threat.

Such drugs may not be new in terms of chemical knowledge or invention, rather they have only recently appeared in the drug market. Often, this happens because there is an attempt to evade the drug control legislation. In the UK, an attempt to control drugs designed to evade such laws was made through the Psychoactive Substances Act 2016 where offences related to anything that is psychotropic are controlled. This meant that substances such as alcohol, caffeine and nicotine had to be specifically exempted. A number of practical and legislative problems around this legislation have been raised and are discussed elsewhere,[4] but in summary the problems are around establishing what the pharmacology of a recreational drug would be without clinical trials. Prior to this, temporary control of some substances was achieved through the use of Temporary Class Drug Orders. Such instruments could only be used to classify a drug for 12 months, during which time information about the drug was gathered and decisions made about permanent control or revocation of the order. At the time of writing, some countries still use similar orders to control newly emerging drugs.

These legislative instruments also lead to the fact that, in addition to the drugs themselves, the drug chemist is also likely to encounter paraphernalia associated with the supply of drugs (for example, scales, knives, and wrapping materials), consumption of drugs [spoons, syringes, smooth-surfaced items, *e.g.* mirrors, or banknotes (for snorting drugs)], or drug residues after the drugs have been consumed. This means that in addition to drug analysis, the drug chemist should be aware of other possible analyses and evidence types in forensic science – for example, toolmarks, the presence of striae in films under polarised light, the chemical analysis of wrapping materials, both for comparative purposes to link drug samples together and the possibility that fingerprints, hairs, fibres and indeed human DNA may be found on drug samples.

1.5 Drug Control in the United States of America

In the United States of America, the controlled substance laws are enforced by the Drug Enforcement Administration. Legislatively, drugs are controlled through the Controlled Substances Act (CSA) 1970 and are listed in five schedules according to their medical use, potential for abuse, and safety or dependence liability (Table 1.2). A number of factors are considered within the CSA when determining in which schedule a substance should be placed. These include

1. the actual or relative potential for abuse of the drug,
2. scientific evidence surrounding the pharmacological effect of the drug, if known,
3. the state of current scientific knowledge regarding the drug or other substances,
4. the history of the drug and its history and current pattern of abuse,
5. the scope, duration, and significance of abuse of the drug,
6. what, if any, risk there is to the public health,
7. the psychic or physiological dependence liability for the drug, and
8. whether the substance is an immediate precursor of a substance already controlled.

In addition to naming drugs within the CSA, the legislation, similar to the UK legislation, controls isomers, esters, ethers, salts, and salts of isomers, esters, and ethers wherever their existence is possible for some drugs and drug types. Thus, there is both specific and generic control of the drug materials. In addition, the CSA controls "materials" and "preparations" that contain some drug materials. Interestingly, the legislation controls some drugs by the mechanism of action/effect – for example, those drugs that

Table 1.2 Scheduling criteria for controlled drugs under the Controlled Substances Act 1971.

	Potential for abuse	Accepted medical use	Potential for addiction
Schedule I	High	None	The drug is not safe to use even under medical supervision
Schedule II	High	Yes, sometimes allowed only with "severe restrictions"	Abusing the drug can cause severe physical and mental addiction
Schedule III	Medium	Yes	Abusing the drug can cause severe mental addiction or moderate physical addiction
Schedule IV	Moderate	Yes	Abusing the drug may lead to moderate mental or physical addiction
Schedule V	Lowest	Yes	Abusing the drug may lead to mild mental or physical addiction

are cannabinomimetic [the synthetic cannabinoid receptor agonists (SCRAs)] and react with a specific receptor in cells, the cannabinoid receptor type 1 (CB-1). This might be considered parallel legislation to that of the Psychoactive Substances Act 2016 in the United Kingdom but the UK legislation is far more generic. The US legislation also outlines the assays and studies in which the action with this receptor needs to be observed.

1.6 Drug Control in Australia

Drug control in Australia is complicated, with state, federal and territory-based control of drugs. Control itself is enacted through the Poisons Standard 2019, which is a compilation of the decisions made under the Therapeutic Goods Act 1989. Schedule 8 of the standards lists those drugs for which possession without a licence is an offence. Schedule 9 lists those substances that are prohibited.

1.7 Drug Control in India

Drugs in India are controlled by the Narcotic Drugs and Psychotropic Substances Act 1985 and its subsequent amendments, and drug trafficking is controlled by the Prevention of Illicit Trafficking in Narcotic Drugs and Psychotropic Substances Act 1988. As with legislative instruments in other countries, drugs are controlled by their chemical names, non-proprietary names and colloquial names where they are widely used – for example, "ganja" and "heroin".

1.8 Conclusion

The drug chemist is faced with a number of analytical challenges. The question(s) asked by the judiciary and the legislative framework in which they work determines the type of analysis to be undertaken. It also determines how the chemist reports their findings, which must be articulated using the legislative language appropriate to the jurisdiction in which the work is being carried out and reported. Additionally, the drug chemist needs to be aware of a wide variety of evidence types that they are likely to encounter and to understand the analytical approaches that can be taken to maximise the evidential value of each.

References

1. Available from: https://theconversation.com/kush-what-is-this-dangerous-new-west-african-drug-that-supposedly-contains-human-bones-220608. Accessed 25 Apr 2024.
2. P. M. Mthembi, E. M. Mwenesongole and M. D. Cole, *Emerg. Trends Drugs Addict. Health*, 2024, **4**, 100142.
3. The Misuse of Drugs Act, 1971, Available from: www.legislation.gov.uk/ukpga/1971/38/contents. Accessed 25 Apr 2024.
4. L. A. King, in *Forensic Chemistry of Substance Misuse: A Guide to Drug Control*, Royal Society of Chemistry, 2022, pp. 48–50.

2 Drug Sampling

2.1 Introduction

The early part of the forensic examination of a drug sample is critical if meaningful conclusions about the samples being analysed are to be drawn. Before any analysis, the forensic science question(s) being asked should be considered. There might be several interrelated questions in a drug case. Questions that are asked might include:

1. is there/are there a/any controlled substance(s) present?
2. how much controlled substance(s) is/are present?
3. if there are controlled substances present, are the samples related in some way?
4. is there any other trace, chemical or biological evidence to relate the samples to other cases?
5. is the amount of drug present typical of simple possession or is it indicative of drug supply?
6. are these controlled substances related to those in other cases?
7. how were these controlled substances made?
8. how can the analysis of any drugs and related materials in the drug samples help establish the distribution routes and networks of the controlled substances in support of intelligence gathering and court-going law enforcement?

To answer these questions, the drug samples require different treatments and analyses. Without these treatments being carried out correctly, the right answers may not be obtained and the correct conclusions may not be drawn. The first step in answering any of these questions is the physical examination and sampling of the

---

Controlled Drug Analysis
By Michael D. Cole, Lata Gautam and Agatha Grela
© Michael D. Cole, Lata Gautam and Agatha Grela 2025
Published by the Royal Society of Chemistry, www.rsc.org

items received for analysis. At this stage, it will not be known, unless an on-site or preliminary screening test has taken place, whether or not there might be controlled drug(s) present in a sample. Prior to beginning any examination of case materials, questions such as:

1. how do I make sure that any controlled drugs that I find come from the sample and from nowhere else?
2. how am I going to test the sample, using the most appropriate methods, to determine whether or not any controlled substance is present, or otherwise?
3. have I got enough sample to answer the question being asked?
4. how am I going to obtain any samples from the materials with which I am presented in order to carry out the relevant analysis(es)?
5. can I leave enough samples for another analyst to examine the material to answer the same or different set of questions?
6. how am I going to ensure that the samples I obtain are suitable for analysis?

should be considered. There is no simple or single answer to these questions. However, by considering the factors described in this chapter and later in this book, it should be possible to obtain a sample that is suitable for analysis and address the questions posed above. It should then be possible to successfully carry out the analyses described in the following chapters of this book.

2.2 Physical Description

The first step in any forensic analysis of a drug sample is a physical examination and description of the item as received in the laboratory in its original packaging. Before any examination takes place, it is especially important to ensure that anybody handling the item being examined and that any workspace, equipment being used, chemicals, reagents and solutions being employed, and storage containers, are free from any controlled substances. This is especially important in the analysis of trace samples. Trace samples can be conveniently thought of as samples that could potentially be easily contaminated, whereas bulk samples might be considered to be samples that are sufficiently large that they could, potentially, contaminate the analyst, the laboratory or other samples being analysed. Ensuring that analysts are free from contamination prior to beginning work can be achieved by taking appropriate swabs (usually the hands and wrists) of anyone handling the sample, swabs of the workspace and samples of chemicals and solvents to be used during the analysis and analysing them for the presence of drugs prior to opening the item being examined. For some compounds [*e.g.* lysergic acid diethylamide (LSD)], there are also recommended post-analysis procedures for testing the workbench (*e.g.* UV light for LSD sample processing) to ensure a clean working space or detect any contamination. However, these procedures would normally be incorporated into the standard operating procedure for the laboratory. For bulk samples, it is sufficient to use certifiably clean personal protective equipment (PPE), laboratory protocols for cleanliness and certified reagents and solutions.

The first stage in the description should be a recording of the exhibit label, including case identifiers, names, dates, the charges being brought (this often tells the analyst the question being asked) and a description of the item contained within the exhibit or production. Further, and importantly, care should be taken to ensure that the continuity of evidence information is complete. If the label is incomplete or continuity of evidence cannot be established, then the analysis should not proceed, because it cannot be determined where any controlled substance identified within the sample actually came from.

Having ensured that all of the information on the label or packaging has been recorded, the actual packaging itself should be described. This should include a statement on whether or not the packaging was intact. This is extremely important, especially in the case of trace sample analysis, because it must be demonstrated that any drug found comes from the sample and no other source. If the packaging has been damaged, this is not, of course, possible and usually the laboratory analysis would not normally proceed.

Having described the item, it should be opened according to standard laboratory protocols, and once the item(s) has(ve) been removed, it/they should be fully described and recorded. This may include one or more of making a drawing of the item, taking a photograph, weighing the item and separating out different items contained within the packaging if appropriate. Where drawings are made, the scale should be depicted, and when photographs are taken, a scale object should be included.

2.3 Trace or Bulk: What Should Be Done?

Having made a physical description of the item and recorded the information on the labels or packaging, the next step in the analysis is to decide whether the sample is a "trace" sample or a "bulk" sample. The reason for this is that these sample types are treated differently, as shown in Figure 2.1.

To make a decision on this, it is worth considering what a trace sample and what a bulk sample might be. Trace items may include, for example, items on which drugs were mixed, the surfaces of knives or items used to weigh drugs, paraphernalia associated with the use of drugs, or perhaps bank notes that are used either for the consumption of drugs or in transactions related to drugs. Bulk samples may include powders that may range in weight from perhaps a few hundred milligrams to tens or hundreds of kilos, liquids that may have been seized, for example, from a clandestine laboratory, or large numbers of tabletted dose forms.

2.3.1 Recovering Drugs from Trace Samples

The way in which a trace sample is usually analysed involves taking a swab or a small amount of material from the item being examined in a way that will allow subsequent instrumental analysis. For example, if the item is a glass surface upon which heroin or cocaine may have been mixed, a cotton wool swab may be soaked in a suitable solvent, depending on the drug thought to be present, and then used to swab some but not all of the surface on which the drugs are thought to have been placed. It is very important that the correct solvent is used so that the drug samples do not break down or react with the solvent itself – the item being examined will often give a clue as to which drug(s) might be present and which solvent should be used. Examples of suitable solvents for swabbing are listed in Table 2.1, and exemplar breakdown products are shown in Table 2.2. When drugs are found in combination, for example, Nyaope or cocaine mixed with piperazines, the problem becomes more complicated but can usually be resolved with thought process and by considering the chemistry of the potential drugs involved. For example, in Nyaope, both cannabis and heroin are known to occur. Cannabis is acid-labile and therefore cannot be dissolved in chloroform, which is a good solvent for dissolving heroin samples. Conversely, cannabis is easily dissolved and stable in low-molecular weight alcohols,

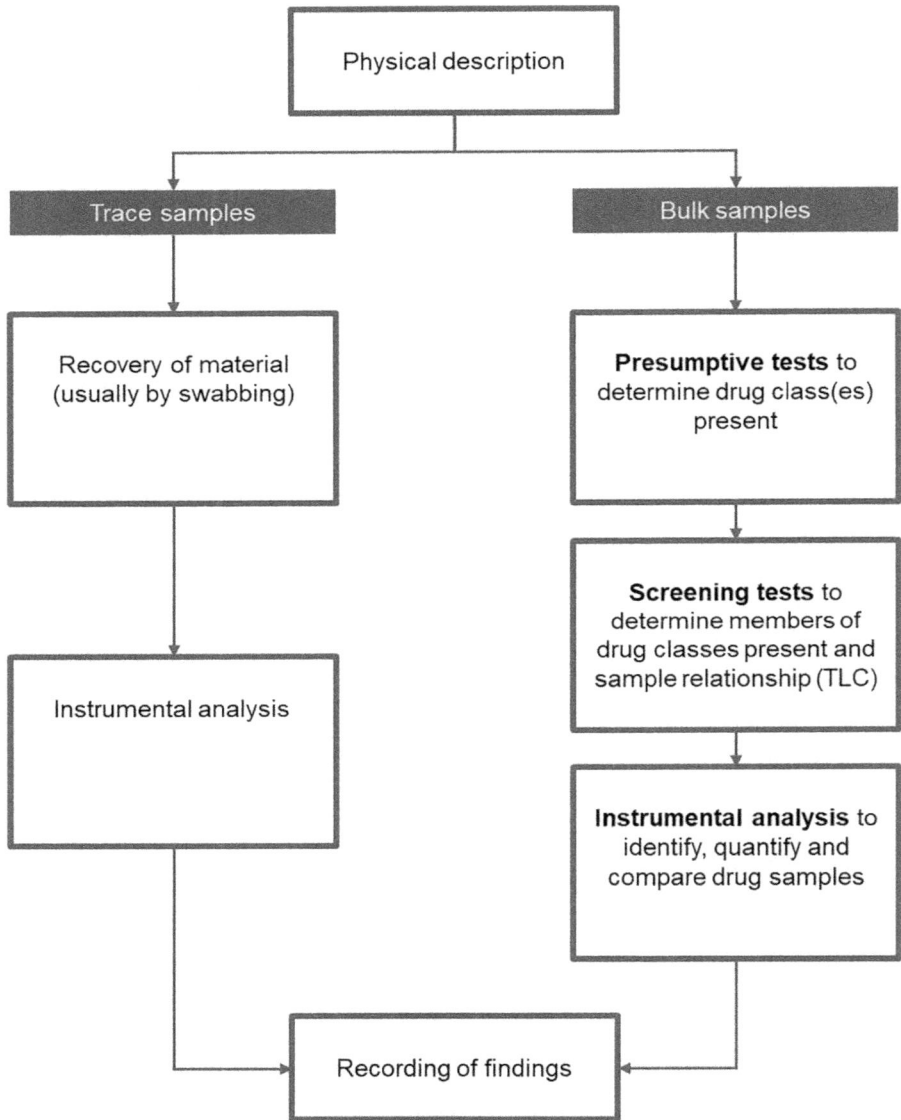

Figure 2.1 Scheme of work for the analysis of trace and bulk drug samples.

but these are hygroscopic and the water they contain leads to the hydrolysis of heroin samples. The best type of solvent to use for drugs of this type is a tertiary alcohol, which does not react with either drug group but will dissolve both drug groups quantitatively.[1] It should also be remembered that a negative control swab should

Table 2.1 Exemplar solvents for the dissolution of drugs prior to chemical analysis.

Drug or drug mixture	Example of unsuitable solvent	Suitable solvent
Heroin	Methanol – contains water, causing hydrolysis	Chloroform
Cocaine	Methanol – contains water, causing hydrolysis	Chloroform
Cannabis	Chloroform – cannabinoids labile in HCl due to the decomposition of chloroform	Ethanol
Nyaope (a mixture of cannabis and heroin)	Methanol, ethanol and chloroform – cause hydrolysis and acid lability	Tertiary alcohol – for example, *t*-BuOH
Piperazines	Dichloromethane – causes the breakdown of fluorinated piperazines	Tertiary alcohol – for example, *t*-BuOH

Table 2.2 Examples of breakdown products of drugs resulting from dissolution in the incorrect solvents.

Drug	Breakdown product	Cause
Tetrahydrocannabinol	Cannabinol	Oxidation
Diamorphine	6-*O*-monoacetylmorphine	Hydrolysis in solvents containing water
—	3-*O*-monoacetylmorphine	Hydrolysis in solvents containing water
—	Morphine	Hydrolysis in solvents containing water

be made using the solvent alone, and a "blank" swab with no solvent should also be analysed to demonstrate that the batch of swabs being used is clean and free of any drugs. If the sample is to be analysed immediately, the swab should be extracted, ideally using the same batch of solvent as that used to take the swab; any solids removed by microcentrifugation, the extract concentrated if required under a stream of dry nitrogen, and analysed using an appropriate instrumental technique. Dry nitrogen is used to concentrate the sample because nitrogen itself is inert, whereas air, for example, contains oxygen, which may oxidise the sample, and being dry will not contain water, which may cause the hydrolysis of some drug types.

If the sample is not going to be used immediately, it should be dried under a stream of dry nitrogen and stored in an airtight container, preferably under an atmosphere of nitrogen, in a dry, dark place. The reason for storing under nitrogen is that this will reduce the risk of any drugs in the sample decomposing by oxidation or hydrolysis. Storing the sample in the dark reduces the risk of degradation of any drugs present due to the ultraviolet light present in daylight. A commonly encountered example of this is the photo-oxidation of cannabinoids in cannabis samples.

2.3.2 Treatment of Bulk Samples

If the samples are bulk samples, a different route of analysis is undertaken. The samples are divided into visually identical groups and then the number of items to be taken from each group needs to be identified. How the sample size is determined is discussed in Section 2.5. The methodology used will depend on the legislative requirements of the legal system in which the analyst is working. However, whichever method is used for determining the number of items to be examined from each group, the items must be chosen at random. In these circumstances, random selection refers to the fact that each item has an equal chance of being chosen for analysis. This can be achieved by assigning each item a number and then choosing the appropriate number of samples using a random number generator or random number tables. Once the items have been chosen for analysis, they are usually subjected to screening tests, sometimes followed by thin layer chromatography, depending on the question being asked, and then one or more instrumental analyses in order to identify the drug, quantify the drug, determine the impurities in the drug and compare one or more drug samples to each other to identify the route by which the drug was manufactured. What is done depends on the question(s) having been asked, the amount of drug available to the analyst and the instrumental methodologies available.

2.3.3 Other Considerations

In addition to any drugs that might be in the sample, there may be other important evidence associated with the materials received for examination. These may include, for example, fingerprints, the physical fit or colour, measurements of wrapping materials, or the striations that can be seen under polarised light on, for example, cling film sheets, which may have been used to wrap the drugs.

The presence of these different evidence types may determine the type of examination and the order in which the examination is performed. For example, if fingerprints are found on wrapping materials or drug-taking paraphernalia, they should be recovered and recorded before any sampling technique is applied to recover drugs present on the surface of the item being analysed. In this way, both the fingerprint and any drug evidence will be recovered.

When considering wrapping materials, it may be necessary to consider whether or not the wrapping materials fit together – *i.e.* whether or not there is a physical fit. For this reason, if an item is thought to contain drugs and needs to be unwrapped, it should be done extremely carefully so that any physical evidence offered by the wrapping material itself is maintained. Going further, if the wrapping material is made of, for example, cling film, it may be that striations that can be seen under polarised light can be used to relate the sample to a roll of film seized elsewhere. Sometimes, the wrapping materials are knotted – especially when made of cling film or placed in balloons. Details of the knot should be recorded as this may allow comparison with other drug samples in the same or different cases, and some drug traders use knots as "trademarks" to identify, for example, the supplier, or the drug content of the powder. When accessing drugs in knotted samples, the knot should not be undone because this may serve as evidence at a later date.

Consideration should also be given to the physical relationship of wraps to each other and to items found elsewhere. For example, pieces of cling film, paper or aluminium foil used to wrap the drugs may contain physical information that relates them to the items found elsewhere, and wherever possible, this information should be recorded and preserved.

In addition, when a sample is being swabbed, wherever possible, the solvent used should be one that does not dissolve or damage the item in question. For example, some plastic films can be dissolved by chloroform, which is preferentially used to recover some drug classes. Under such circumstances, another solvent should be used, and if it is likely that the solvent may lead to the decomposition of the drugs, which are likely to be trace samples, the extract should be dried immediately under nitrogen and either stored or re-suspended in a solvent that is not thought to cause the decomposition of the drug class in question and then analysed. If in doubt, experiments should be conducted using different solvents on different makes or brands of, for example, plastic films in this case, to determine whether or not they are solvent soluble prior to attempting to recover a sample from the case materials. It is imperative in forensic science that samples are not taken without the knowledge that the case samples will not be irreversibly damaged.

2.4 Packaging of Drug Materials

All drug materials should be received in the laboratory in a condition that the analyst can confirm that the packaging containing the drug is intact. The type of packaging used will depend on the drug material submitted for examination. It will also depend on the question that is going to be asked of the analyst. Fresh plant or fungal material is particularly problematic because if it is not stored properly, there is a risk that it will become mouldy, making it difficult or impossible to analyse. Fresh material should be packaged in paper bags, which are porous and allow the sample to "breathe" and dry out, and stored at room temperature. Plastic bags should not be used because the moisture in the sample cannot escape. Ideally, the samples should also be stored in the dark to minimise the risk of photo-oxidation.

Resinous materials, which may have physical evidence as well as chemical evidence on the item, should be packaged in such a way that the physical evidence is not damaged. Such evidence includes edges where blocks of resin may fit together, scratches or striations on the surfaces of the resin that may be used to link one sample to another, knife or tool marks where tools have been used to divide the resin up, and marks and striations caused by the wrapping material. It may be sufficient that the item is contained within an exhibit bag, but if the physical evidence is particularly important, it may be worth considering packaging the item in a box to protect it.

The packaging of powdered samples depends on the question that is going to be asked in terms of analysis. If a straightforward identification of a drug is required, then it is possible to pack the material in, for example, a polythene bag or an exhibit bag. However, there is significant risk that the drug will form an aerosol when removed from such packaging and great care should be taken to avoid laboratory contamination.[2,3] If, however, an analysis of occluded solvents (solvents trapped in the crystalline structure of the drug) in the sample is to be carried out, such packaging is insufficient because it is permeable to organic solvents. When analysis of occluded solvents is to be undertaken, it is best that the item is packaged in a glass screw-top bottle with a small headspace. This can prevent the contamination by organic solvents or the escape of solvents from the sample.

Some drug samples might be treated as documents. For example, street samples of LSD blotter acids are frequently manufactured with the drug impregnated on paper. The analysis type that is to be undertaken is not dissimilar, in part, to document examination. It is, therefore, perhaps sensible to use a plastic or evidence bag or, when convenient, a box so that the physical evidence available on the item is not damaged. LSD is also an example of a drug where the packaging should protect those handling it. LSD is one of the most potent hallucinogens known to man and is readily absorbed through the skin. Any packaging for this type of drug should be made certain to contain the drug within the package for the safety of all concerned.

It is particularly important that liquid samples are correctly packaged so that they do not leak or evaporate. Ideally, the whole container for a liquid sample should be taken back to the laboratory for analysis. However, sometimes this is not possible, and the sample should be taken from the bulk using, for example, a glass pipette. The liquid should then be placed into a screw-top vial from which it cannot leak or the vapour cannot escape. For such samples, it is especially important that the contamination and changes to the sample are considered. It is for this reason that glass is preferred to plastic. The problem with plastic is that the plasticisers may leak into the sample or that the drugs or impurities (which may be relevant for drug comparison and profiling) may sorb onto the plastic itself.

Samples that have been taken with a swab are easily contaminated and special care should be taken to protect them. Where they are taken at a scene, they should then be placed within a container that will protect them. If they are not to be analysed immediately upon return to the laboratory, they should be dried to prevent decomposition of the drug sample and, where possible, stored in the dark in order that photo-oxidation of the sample is prevented.

2.5 The Number of Samples to Take

When the analyst has a large number of items that must be analysed, it is sometimes necessary to make decisions about the number of items that should be analysed. This, more often than not, applies to bulk samples but can also occur when the analyst encounters a very large number of trace samples. There are several approaches that can be adopted but there are also some underlying principles that must be recognised and adhered to.[4]

The first step in this process is to divide the items into visually identical groups. These will then form the populations from which the samples will be taken. Two guiding principles must be adhered to at all times when taking samples from these populations. The first is that whatever the sample size is, the samples must represent the population as a whole. The second is that the sample taken from the population must be taken at random. While it is possible to use random number generators for small- to medium-sized populations, for very large-sized population, the method recommended by the European Network of Forensic Science Institutes Working Group on Drugs is the "black box" approach. In this approach, the items are placed inside a black bag or box and the samples are drawn from these items. While this is not a truly random sampling technique, it does at least provide some assurance that a representative sample has been obtained.

There are a number of methodologies that have been applied for determining the sample size which should be taken from a population of items being examined. The first of these is a set of techniques that involves sampling in arbitrary numbers and the second is statistical methods based on a number of assumptions about the drug population.

Table 2.3 Advantages and disadvantages of arbitrary sampling regimes for the determination of sample size ahead of drug analysis.[4]

Sampling technique	Advantages	Disadvantages
Sample all ($n = N$)	100% of the population sampled; easy to explain	Very large sample sizes possibly resulting in costly and lengthy analysis
$n = 0.05N$, $n = 0.1N$ etc.	Simple approach; easy to explain	Very large sample sizes possibly resulting in costly and lengthy analysis
$n = \sqrt{N}$, $n = 0.5\sqrt{N}$ etc.	Frequently used; relatively easy to explain	When N is small, the sample size may be very small; when N is large, large sample sizes are possible
$N = 20 + 10\%(N - 20)$	Likely to discover any heterogeneity in the sample	Requires population >20; large sample sizes possibly resulting in costly and lengthy analysis
When $N < 10$, sample all When $10 \leq N \leq 100$ sample 10 $N > 100$ sample $n = \sqrt{N}$	Recommended by the United Nations Drug Control Programme	When N is large, a large sample is possible; difficult to explain to a "lay" audience

2.5.1 Arbitrary Sampling

Several arbitrary techniques have been applied to drug sampling. They each have advantages and disadvantages,[4] which are given in Table 2.3. Another approach has been to limit the sample size (n) to three items (3) and this has been adopted by some jurisdictions around the world.[5] The method used will depend on several factors, including the requirements of the jurisdiction in which the work is being carried out and the question being asked. If the question is "Is there a controlled drug present?" it may only be necessary to sample and analyse one item if a controlled drug is found. If the question is about the amount of drug present, the proportion of items that contain controlled substances, or whether items are related, then a larger sample size is needed.[6] In short, there is no "right" answer to how many items should be sampled, but there is a wrong way to do it. For example, if an estimate of drug content is required, a sample size of one is not appropriate unless the seizure is a single item.

2.5.2 Statistical Sampling

The most commonly used approach in statistical sampling to identify whether or not controlled substances are present is the hypergeometric sampling technique. The assumption made is that a

Table 2.4 Number of samples (n) to be taken from a population of size N with a 95% probability that 90% of the items contain the drug of interest according to the hypergeometric distribution.[4]

Number of items (N)	Number to be tested (n)
10–12	9
13	10
14	11
15–16	12
17	13
18	14
19–24	15
25–26	16
27	17
28–35	18
38–46	20
59–77	23
89–118	25
179–298	27
>1600	29

fixed but unknown proportion of items contains drugs. No assumption is made about homogeneity. On this basis, the proportion of items containing drugs can be identified as the probability that they contain drugs. The number of items to be sampled randomly from the population can then be identified. This is exemplified in Table 2.4. In this example, there is a 95% probability that 90% of the items in a population of size N contains the identified compound and the sample size to be obtained is given.[4]

2.6 Representative Sampling of Drugs

Having determined the number of items to be examined, and selected them, it is then necessary to ensure that samples are homogeneous prior to their analysis. This is to ensure that the materials that are analysed represent the item from which they have been taken. This can present particular problems, especially when the samples are very large. Depending on whether the sample is a powder, a liquid or a tablet, a slightly different protocol applies.

2.6.1 Sampling of Powders

When a large amount of powder is present, it can be difficult to obtain a representative sample. Examples might include dustbin bags containing large amounts of heroin powder. Under such circumstances, the usual method is to take one or more "cores" of the material using a glass tube. This will provide a sub-sample of the material to be examined to which methods for smaller amounts of powder can be applied.

The cone and square method is recommended by the United Nations Office on Drugs and Crime (UNODC) for small- and medium-sized samples. An amount of powder is placed on a porcelain tile and homogenised with a spatula or a glass rod. The material is then divided into quarters. Two of the quarters are returned to the original packaging and the remaining material is combined. This process is repeated until a small sample representing the item is obtained. The size of this sample can vary but is typically between 10 mg and 100 mg.

An alternative method that is available, for samples of sufficient size, is the use of a blender. The problem with such a method is that drug powders can get underneath the blades and also into the sealing mechanism of the blender. It is therefore doubly important that the blender is demonstrably clean prior to any homogenisation so that the source of any drug found can be clearly identified.

Plastic bags have also been used to homogenise powdered drug samples. The powder is placed in the bag, and the bag is shaken vigorously. The principal problem of this methodology, other than the bag itself breaking, is that sometimes the drugs carry a charge and stick to the inside of the plastic bag. Therefore, the amount of drug measured in the sample will be low and unrepresentative of the true value. Drug powder can also escape from the bag and contaminate the laboratory.

Some drug samples comprise more than one type of material; such a sample is Nyaope, a drug found in South Africa (Figure 2.2). This is a complex mixture of heroin (powder), cannabis (plant material including hard seeds), antiretrovirals (powder), tobacco (plant material) and other excipients (usually powder).[1] Such samples may require grinding, for example, in a rock mill, prior to laboratory homogenisation and analysis. This may be the only way to ensure that the sample is homogeneous.

2.6.2 Sampling of Liquids

Liquids should be sampled in such a way that there is enough liquid remaining for subsequent analyses. It is assumed that any particular

Figure 2.2 An example of a street sample of Nyaope from South Africa. The mixture is a combination of tobacco, cannabis herbal material and seeds, powdered heroin and antiretroviral compounds, for example, lopinavir and efavirenz.

phase of a liquid is homogeneous. Sometimes, if there are precipitates or emulsions, this may not be true and attempts should also be made to collect this material. The liquid should be collected using, for example, a glass pipette and the material collected should be placed into as small a glass vessel as possible and sealed with an airtight seal. It is important that the headspace is minimised so that the possibility of oxidation and/or evaporation of the sample is reduced.

2.6.3 Sampling of Tabletted Dose Forms

Tabletted dose forms should be very carefully sampled. Quite often, there will be manufacturing marks that are either deliberate or accidental on the surface of the tablet. These marks are known as "ballistics".[7] The assumption made of a tablet is that the material contained within it is homogeneous. On this basis, the sample should be taken away from any ballistics that are found on the surface of the tablet and should come from the central portion of the tablet so that the bulk of the material is represented. The tablet should not be completely ground up, and where at all possible, it should not be broken. Again, the powder which is obtained should be placed within as small a container as possible to reduce the risk of oxidation.

2.6.4 Sample Dissolution

Once the samples to be analysed have been identified, it is often necessary to dissolve them in a solvent in preparation for the next analytical step. These solvents must satisfy several basic criteria. Firstly, they must dissolve the drugs quantitatively over a wide concentration range. They must not react with the drugs during any stage of the extraction process or sample preparation. They must not cause the drugs to break down. They should not present significant health and safety risks and should be easily disposed of. They must also be cost-effective because it is likely that large volumes in operational laboratories will be used. Finally, but not insignificantly, they should be compatible with the analytical techniques used on any particular drug group. Once the casework material has been described, sampled and, where necessary, dissolved, the next steps in the analysis can proceed.

References

1. P. M. Mthembi, E. M. Mwenesongole and M. D. Cole, *S. Afr. J. Sci.*, 2021, **117**(11–12), 1.
2. E. Sisco, E. L. Robinson, R. Mead and I. V. Miller, Measuring changes in drug particulate on evidence packaging due to routine case analysis, *Forensic Chem.*, 2021, **26**, 100372.
3. E. Sisco, M. Najarro and A. Burns, A snapshot of drug background levels on surfaces in A forensic laboratory, *Forensic Chem.*, 2018, **11**, 47.
4. S. Schiavone, M. Perrin, H. Coyle, H. Huizer, A. Block and B. Cardinetti, in *Guidelines on representative drug sampling*, European Network of Forensic Science Institutes, 2003, p. 56.
5. C. G. G. Aitken, Sampling - how big a sample?, *J. Forensic Sci.*, 1999, **44**, 750.

6. T. Csesztregi, M. Bovens, L. Dujourdy, A. Franc and J. Nagy, Sampling of illicit drugs for quantitative analysis – Part III: Sample plans and sample preparation, *Forensic Sci. Int.*, 2014, **241**, 212.
7. C. Zingg, *The analysis of ecstasy tablets in a forensic drug intelligence perspective*, Doctoral dissertation, Université de Lausanne, Faculté de droit et des sciences criminelles, Switzerland, 2005.

3 The Analysis of Cannabis and Products

3.1 Introduction

3.1.1 Cannabis

Cannabis, in locations where it is a controlled drug, is the most commonly encountered drug in the forensic science laboratory. All products are derived from the plant *Cannabis sativa* L. Drug materials can occur in a wide variety of forms including those derived from plant material, resinous material, or liquids. To give a sense of scale of the use of cannabis, in 2018, 5610 tons of cannabis products (4303 tons of herbal material and 1307 tons of resin) were seized globally with there being an estimated 192 million users.[1] Data from the European Monitoring Centre for Drugs and Drug Addiction suggest that 74% of all drug seizures in Europe are related to herbal cannabis, cannabis resin or cannabis plants,[2] but there are changes in the materials being seized. Between 2009 and 2019, in Europe, the quantity of cannabis resin seized has remained relatively constant, while the amount of herbal material seized has increased by approximately 240%.[2]

The active drug in cannabis products is Δ^9-tetrahydrocannabinol (Δ^9-THC), which is approximately 100 times more active than the positional isomer Δ^8-tetrahydrocannabinol (Δ^8-THC) (Figure 3.1). The Δ^9-THC is derived from its natural precursor cannabidiol (CBD) and oxidises to cannabinol (CBN). While there are in excess of around 100 known cannabinoids, these four are the ones of principal concern to the forensic scientists in addition to the closely

---

Controlled Drug Analysis
By Michael D. Cole, Lata Gautam and Agatha Grela
© Michael D. Cole, Lata Gautam and Agatha Grela 2025
Published by the Royal Society of Chemistry, www.rsc.org

related Δ⁹-tetrahydrocannabinolic acid (Δ⁹-THC-COOH, THCA). This latter thermally decarboxylates to Δ⁹-THC when cannabis material is smoked. This is of relevance later in this chapter when analytical methodologies are considered.

If cannabinol is identified in a cannabis sample, the identification of the drug can still be made, while the sample cannot be used for chemical profiling or comparison purposes because the original composition of the drug will not be known – as it has started to break down. Cannabinol is the key marker for this decomposition, and thin layer chromatography (TLC) can be used at an early stage

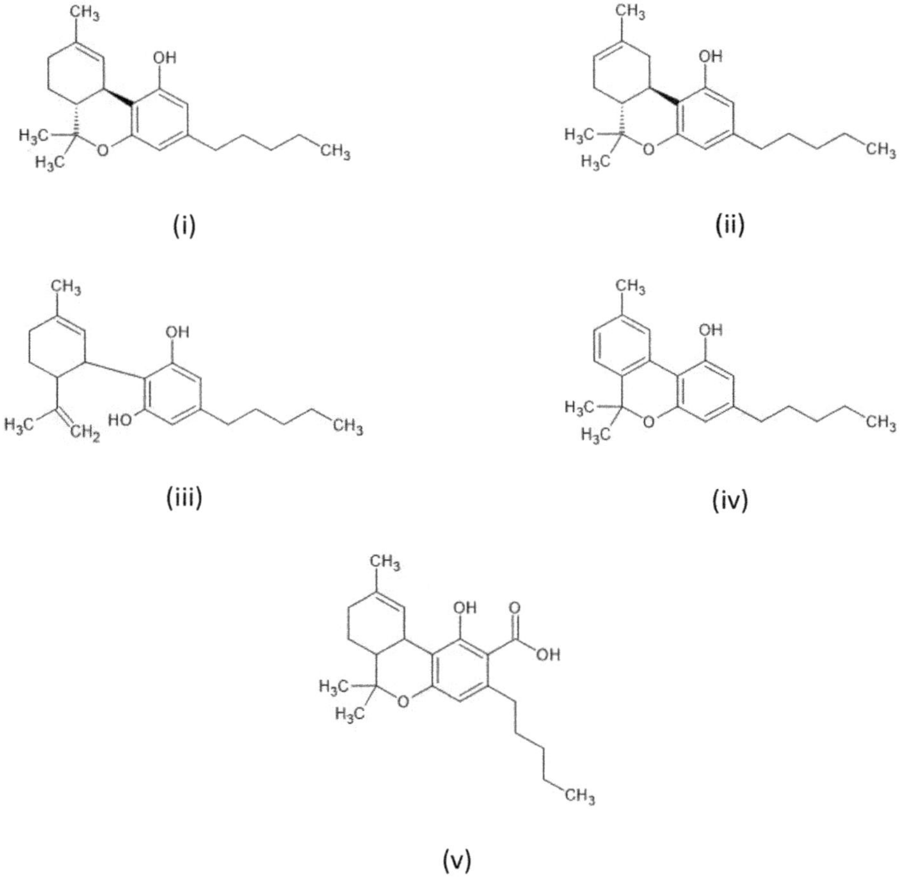

Figure 3.1 Structure of (i) Δ⁹-tetrahydrocannabinol (Δ⁹-THC), (ii) Δ⁸-tetrahydrocannabinol (Δ⁸-THC), (iii) cannabidiol (CBD), (iv) cannabinol (CBN) and (v) Δ⁹-tetrahydrocannabinolic acid (Δ⁹-THC-COOH, THCA).

in cannabis analysis to determine whether or not oxidation products are present. Where they are, this does not, however, preclude the sample being used for comparison where a physical fit between two resin samples can be established.

Cannabis can also be mixed with other drugs. Examples include a South African drug mixture known as White Pipe where cannabis is mixed with methaqualone (the latter is also known as Mandrax) and smoked in the top of a broken glass bottle. It may also be found in drugs such as Nyaope, which is a mixture of cannabis, heroin, anti-HIV antiretrovirals (for example, efavirenz and lopinavir), and tobacco, a mixture that is commonly smoked in cigarettes.[3]

3.2 Analytical Sequence for the Analysis of Cannabis

The most frequent way that cannabis is administered is by smoking or by eating a material containing the drug. The latter are also known colloquially as "edibles". This means that as well as the drug containing materials themselves it is possible that casework materials may include paraphernalia associated with drug taking, or, indeed, used to wrap the drug. Such items may include unfinished cigarettes containing drug material, residues in, for example, ash trays, knives and items used to divide blocks of resin, scales that have been used to weigh the cannabis materials and paper, plastic film or other materials used to wrap the items. Items seized may also

Figure 3.2 Examples of cannabis "edibles".

include edible items ("edibles"), including cakes, biscuits and other items (Figure 3.2).

3.2.1 Bulk *vs.* Trace Samples in the Analysis of Cannabis Samples

In general, drug materials encountered in cannabis samples are bulk samples – that is they are not easily contaminated. However, the paraphernalia associated with the distribution, preparation and use of cannabis may have only trace amounts of the drug present that may not even be visible to the naked eye. Under such circumstances, the visual record of the items in question should be made, and following this, small samples should be taken using an ethanol swab, remembering the need to have material in case a second or subsequent analysis is required. Ethanol is used because of the free solubility of the cannabinoid drugs in this solvent but does not catalyse their breakdown. The swab should then be analysed immediately using the gas chromatography–mass spectrometry (GC–MS) method described in Section 3.2.4.4 or dried under nitrogen and stored (under an atmosphere of dry nitrogen) until analysis.

3.2.2 Physical Description of Cannabis-containing Materials and Associated Paraphernalia

A physical description of cannabis samples can tell you a great deal about the drug that you are examining including possible geographic location of the origin of the cannabis. For example, a resin that is friable and yellow and found in tiles may have originated from the North African coast. Cannabis resin that is sticky, dark and resinous is likely to have originated on the Indian subcontinent. A comprehensive set of physical characteristics relating to the origin of the drug has been compiled,[4] and while these characteristics do not prove the origin of the drug, they do provide features that might be used to establish relationships between samples. It is also possible to establish relationships between samples when, for example, resin blocks fit together or, alternatively, the wrapping materials themselves fit together. Additionally, tools that may have been used to cut cannabis resin, for example, may leave tool marks and striations on the resin surface, which may match a drug sample to a tool or drug samples to each other. Finally, it may be possible to establish relationships between, for example, plastic film rolls used to wrap cannabis, by viewing the striation lines under polarised light. If the birefringence lines match, then it can be opined that the different wrappings came from the same roll of film. It may also be possible to establish relationships between wrapping materials that do not necessarily fit together. All of these highlight the importance of carrying out a good physical description of cannabis samples ahead of any other analyses. Indeed, if pieces of cannabis resin fit together, it is not necessary to carry out extensive and expensive chemical experiments to establish a relationship between the items – but it will be necessary to demonstrate the presence of cannabinoids in order to draw the conclusion that the material is a cannabis sample. The physical description will also determine whether the material should be treated as a trace or a bulk sample (Section 2.3) and what the analytical sequence should be.

3.2.2.1 Physical Description of Herbal Material

Herbal cannabis material generally comprises leaves, stems, seeds and flowering tops as can be seen in Figure 3.3. It is these latter structures that contain the majority of the drug (Δ^9-THC). The leaves are typically palmate, serrated, arranged in pairs and oppositely on the stem and have an easily recognisable shape and smell.

On some occasions, it is not necessary to use complex chemistry to identify cannabis. Sometimes, when plant material is submitted to the laboratory, it is possible to identify cannabis unambiguously using botanical features. Cannabis has three types of hairs, known as trichomes, on its surface, as shown in Figure 3.4. The first type of trichome is the glandular trichome – it is here that cannabis resin is formed and where the drug is contained. The second type is the cystolithic trichome, which looks like a small hook on the surface of the plant. The third type is the unicellular trichomes that look like hairs on the surface of the plant. Unusually, in cannabis, these hairs would point upwards, up the plant, in one direction.

Cannabis is unusual in that it is the only plant in the world that has this combination of hairs or trichomes on its surface. If all three can be identified under the microscope, then it is possible to say that the sample is unambiguously herbal cannabis material. It is sometimes possible to see the remains of these hairs in, for example, ash that has come from a cigarette containing cannabis, as shown in Figure 3.4. However, they are not usually intact and chemistry is needed in order to identify the sample beyond doubt.

Sometimes, living plants are seized. If the plant material is wet, it should be stored in paper bags in a dry, dark environment so that

Figure 3.3 Cannabis material showing palmate, serrated leaves and brown resinous material frequently found in herbal cannabis.

the material can dry out and the plant material going mouldy is prevented as is decomposition of the drugs in the sample by light. Some cases require an estimate of the amount of plant material present in the sample. In order to achieve an estimate of this, the plant material should be dried and either weighed as a whole or a suitable sample taken and the total mass estimated from the sample. Such analyses are required, for example, where a large amount of plant material has been seized and its street value needs to be known or where the charge is drug supply rather than possession (Section 3.2.2.4).

3.2.2.2 Physical Description of Resinous Material

Resinous material may occur in a wide variety of forms ranging from large, complete blocks to smaller, single-dose fragments. Most typically, the resin in street samples is the resin derived from compressed blocks of resin called "soap bars". The resin is collected from the plants, particularly those parts of the plant rich in the glandular trichomes where the cannabis resin, and hence cannabinoids, are formed. These can then be formed into blocks roughly the same size and shape as a bar of soap. Typically, these are dark brown

The Analysis of Cannabis and Products

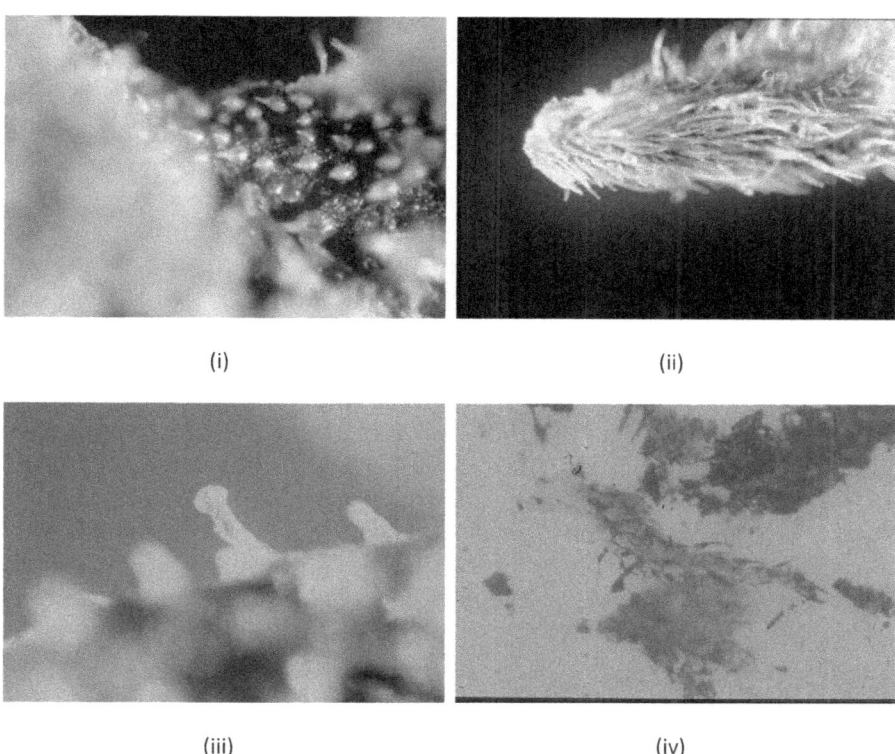

Figure 3.4 Examples of (i) cystolithic trichomes, (ii) unicellular trichomes, (iii) glandular trichomes found on the surface of the herbal cannabis material and (iv) burned trichomes found in cigarette ash after the burning of cannabis material in tobacco.

on the outside and sticky, resulting from the photo-oxidation of the cannabinoids. It is the outside and broken edges of the soap bars where the tool marks and physical fit characteristics are likely to be found, but because of photo-oxidation, the dark material can be used for identification and confirmation of the presence of cannabinoids; however, it does not provide material that can reliably be used in the comparison of cannabis samples. The inside of such soap bars is where the material is more friable, has not been subject to photo-oxidation and it is that which allows a more reliable sample to be obtained for chemical comparison of the cannabis samples. However, chemical comparison may not be needed, even when the chemical nature of the cannabis resin has completely changed. For example, cases have been known where parts of resin have been stored in a toaster, the heat of which will have caused much chemical decomposition of the sample. When compared to another sample that had been stored at room temperature, the two pieces were found to fit exactly together, negating the need for extensive chemical analysis of the samples to establish a relationship between the two.

A good description of the material includes a record of the size, shape, colour, smell and weight of the resin. A note should be made of whether or not blocks of resin in the case, or related cases (where appropriate), fit together. Further records should be made of any physical markings on the surface of the resin, which may become important in linking samples at a later stage. This latter is especially important where a case involves establishing a relationship between drug samples and paraphernalia associated with their preparation. Such marks may include tool marks in the resin left by cutting implements or impression marks left by individual characteristics on weighing implements/devices. Any wrapping material should also be carefully considered, as should the possibility that fingerprints or other trace evidence could be recovered from the sample prior to any destructive chemical analysis.

3.2.2.3 Physical Description of "Hash Oil"

Liquid materials or hash oil is derived from materials that have been extracted from cannabis using petrol. Plant or resinous material is mixed with petrol, in which the cannabinoids, plant oils and waxes dissolve. The solid material is removed and the petrol is evaporated from the sample, leaving a residue that is oily, dark brown and sticky. Such materials are typically dropped onto tobacco and smoked. A physical description of hash oil typically involves recording the colour, smell and volume or mass of the oil present, in addition to recording details of the container in which the oil is found.

3.2.2.4 The Importance of the Weight of the Sample

A part of the physical description of the sample is to establish the sample weight (or volume for hash oil). This is especially important in cases where the charge is to supply cannabis. In cases where it is claimed that the material is for personal use rather than supply, the mass of the material present in the sample under analysis is divided by the assumed mass of plant material used by the users of the drug in a cigarette. For example, a user may use between 100 mg and approximately 1 g resinous material per cigarette. The actual amount used will depend upon the strength of the cannabis and the habituation of the user. The mass of the resin in the seized sample is thus divided by the lowest mass (100 mg), giving the maximum likely number of doses, and the largest (1 g), giving the smallest number. These are then considered in the light of information around the number of times the user takes cannabis in a day. This can be used to estimate the number of days that the material is likely to last and conversations can then occur around whether this is reasonable or not and/or whether the cannabis is likely to decompose in such a timeframe. Using timelines in which cannabis goes off (3 months is a rule of thumb used by some forensic scientists), the length of time that the cannabis will be usable may be estimated.[5] Then the court will be able to make decisions around whether a claim for personal use is reasonable, or whether the mass of material is so large that this is not, and supply is more likely. In cases where value is critical, the same process is used to work out the number of doses and the value is then considered against the relevant street cost of the drug. In all of these scenarios, there is likely to be some error in the estimates and the drug analyst should also make the court aware of such possibilities.

3.2.3 Sampling Cannabis Materials

Once the resin samples have been described and weighed, the next stage is to take a sample for chemical analysis. When sampling plant material, it is important to take samples, where possible, from different parts of the plant, especially when comparisons are being made. For identification purposes, this is less important. The reason that the sample(s) should contain different plant parts is phenotypic plasticity – a phenomenon caused by the effect of the physical environment in which the plant was growing on the cannabinoid content of the cannabis resin.[6] The result of this is that different parts of the same cannabis plant will contain different cannabinoid profiles.

In the case of resin, where the sample is taken from is important because it should avoid all of the physical characteristics of the resin sample and any point on the sample that may interface or fit together with another block of resin. Additionally, if there is a photo-oxidised layer on the outside of the resin, this should be avoided because it will not be representative of the whole bulk sample. In order to take a sample of the resin, a small amount is scraped away from the surface of the block using either a mounted needle or a scalpel and a sample of the main resin is taken. Some laboratories take more than one sample because there has been debate in the literature around the homogeneity of cannabis resin blocks. Once the sample has been taken, it can be analysed for the presence of cannabinoids, the chemical markers for cannabis, by presumptive testing or by chromatographic analysis.

3.2.4 Chemical Analysis of Cannabis

If the plant material cannot be identified using botany (Section 3.2.2.1), and for all samples of resin and hash oil, it is necessary to use chemistry to confirm the identity of the drug. The first stage in this is the use of colour or presumptive tests, then followed by detailed chemical or DNA analysis, depending upon the forensic science question being asked.

3.2.4.1 Presumptive Tests for Cannabis

The tests that are used for the tentative identification of cannabis include the Duquenois–Levine test (one of the most frequently used tests),[7] the Corinth IV test, the furfural test and the alkaline beam test. These are not definitive tests for cannabis products, but tests for the phenolic compounds that are found in cannabis products. It should be noted that a number of materials will give false positives, similar to the results obtained from older cannabis samples with these tests. By way of example, some brands of instant coffee and the herbs mace and thyme, particularly when the samples are fresh, will give a positive colour reaction on the addition of chloroform to the test mixture in the Duquenois–Levine test. However, the rate of colour reaction, the blue colour that is produced and the extraction of the blue complex into the chloroform layer are not the same as those obtained from cannabis. However, the problem lies in the fact that visual colour assessment is subjective and reaction rates are temperature and concentration dependant. It is for this reason that positive, negative and reagent controls should always be undertaken during such analyses, against which the colour reaction and behaviour of the test can be compared.

The Duquenois–Levine test requires three reagents. The first of these is 2.5 ml of acetaldehyde and 2 g of vanillin dissolved in 100 ml of 95% ethanol. The second reagent is concentrated hydrochloric acid and the third reagent is chloroform. The test is performed on a small sample of material in a test tube. Many analysts "wet up" the sample with one drop of ethanol prior to carrying out the test to bring the cannabinoids into solution. The test is performed by adding five drops of the first reagent followed by shaking. Five drops of hydrochloric acid are added and finally 10 drops of chloroform. At this stage, two phases are formed and the presence of cannabis is indicated by the extraction of a blue or purple complex into the lower chloroform layer. Other plant phenolics may also react in the system. However, the blue colour is not the same as that obtained from the positive controls, and sometimes, an emulsion is formed between the aqueous layer and the chloroform layer. Finally, the speed of extraction of the complex formed with the drugs into the chloroform layer is not the same as for the cannabinoids. The extraction of the coloured complex formed when cannabis is present is relatively rapid, while it is much slower in so-called "false positive" reactions.

The Corinth IV salt test is easily performed using a filter paper. A small amount of material under examination is placed on the filter paper and a second filter paper placed on top. Some analysts apply a small amount of pressure to the material in the test to break the test material up and increase the surface area over which the extraction of cannabinoids may happen. On top of the material, on the upper side of the upper filter paper, a few drops of diethyl ether are placed. Diethyl ether is used to extract non-polar cannabinoids but has the disadvantage that it will not extract cannabinol or cannabinolic acid that are too polar to dissolve in the ether. However, diethyl ether has the advantage that it evaporates rapidly, concentrating the sample. The top layer of paper is removed and a small amount of Corinth IV salt in anhydrous sodium sulfate is placed on this paper between the edge of the paper and the dried extract. A drop of water is added outside of this and is allowed to percolate through the paper and through the Corinth IV salt, carrying it towards the cannabis extract. If cannabinoids are present, they will react with the Corinth IV in solution producing yellow, orange, pink and red colours depending upon the composition of the cannabis materials.

When using the furfural test,[7] the material suspected of containing cannabis is ground to a fine powder, placed in a test tube and shaken with 1 ml of petroleum ether. The extract is transferred to an evaporating dish and evaporated to dryness over a steam bath, an operation that should be performed in a fume cupboard. Between three and five drops of ethanol, three drops of furfural reagent [1% w/v furfural (furan-2-carbaldehyde) in ethanol] and 1 or 2 drops of hydrochloric acid are sequentially dropped onto the test substrate.

The mixture is evaporated to dryness and 1 or 2 drops of 55% sulfuric acid in ethanol are dropped onto the mixture. If a purplish red residue forms on the addition of the acid reagent, the sample may contain cannabis.

The alkaline beam test[7] involves placing a small amount of suspected material in a test tube that is shaken with 1 to 2 ml of petroleum ether. The extract is transferred to an evaporating dish and evaporated to dryness over a steam bath. This should be performed in a fume cupboard. To the residue, 1 to 2 drops of the test reagent comprising a 0.5% solution of potassium hydroxide in ethanol should be added. The problem with this test is that the rate of development of the colour reaction, which produces a pink to purple product depending upon the sample, will depend upon the cannabis concentration, the temperature of the room and whether or not there are other materials present in the sample.

3.2.4.2 Thin Layer Chromatography of Cannabis Samples

Once a presumptive test has been performed on the material thought to be cannabis, the next stage may be to identify the cannabinoids present in the sample or for a comparison between samples to be made. One rapid, cost-effective method for doing this is TLC. The advantage that this method has is that it is considerably quicker and less complicated than GC–MS or indeed high-performance liquid chromatography (HPLC).

In order to carry out a TLC analysis of cannabis, the sample should be dissolved in a solvent that freely dissolves the cannabinoids but does not cause them to break down. Ethanol is frequently used for this purpose. The samples should be ground up to a fine homogeneous powder and the solutions are prepared such that the cannabinoid concentrations lie somewhere between 100 and 1000 µl ml^{-1}. In order to achieve this, it is important to have an estimate of the potency (Δ^9-THC content) of cannabis products. Herbal and resinous cannabis potency (Δ^9-THC content) generally ranges from 1 to 16% with some exceptionally potent strains containing up to 30% Δ^9-THC by weight. Concentrations likely to be encountered are best determined at a local level because there is a wide regional variation.

Once the extract is prepared, it should be filtered or preferably centrifuged to remove any solid particulates. Centrifugation avoids the problems of sample loss and/or potential contamination from, for example, the filters or filter paper. Solid particulates will result in poor chromatography and difficulties in comparing cannabis samples or indeed identifying the cannabinoids present. Standard compounds should also be chromatographed, using solutions in the same concentration ranges as those expected from the samples.

Once the solutions of the cannabis extract and cannabinoid standards have been prepared, small amounts of the solutions should be spotted onto the thin layer chromatography plate. Each application should contain between 5 µg and 10 µg of the compounds of interest in order that they are clearly visualised by the spray reagent after the thin layer chromatogram has been developed. The stationary phase is typically silica gel with a fluorescent indicator. The indicator typically fluoresces at 254 nm, meaning visualised spots appear as purple spots on a bright (usually yellow) background. Once the extracts, standard solutions and negative controls have been applied to the chromatography plate, they should be allowed to thoroughly dry and the TLC plate then placed in the chromatographic tank that has previously been saturated with the mobile phase. For cannabis, the mobile phase is typically toluene although other systems have been described.[8] The thin layer chromatogram should be allowed to develop and then removed from the chromatographic tank. The solvent front should be drawn and chromatogram allowed to dry. It should then be visualised under ultraviolet light at 254 nm. Compounds that are present will appear as dark purple spots on a fluorescent background (usually yellow). Any visualised compounds should be recorded either through photographing or drawing the chromatogram. The process should be repeated at 360 nm and again a recording of any compounds that are visualised is made. The chromatographic plate should then be sprayed with a visualising reagent. This is usually 1% fast blue BB in water. A variety of yellow, orange, red and brown compounds form as the reagent reacts with the cannabinoids and other plant phenolics. Again, the chromatographic plate should be drawn and the R_f values of any visualised compounds are calculated.

If, following TLC, cannabinol is found to be present, the sample can still be identified as cannabis. However, conclusions around comparing cannabis samples should be made with extreme caution because cannabinol is one of the breakdown products of Δ^9-THC. The problem with the presence of a breakdown product is that the rate of decomposition under the specific conditions in which the sample has been stored is not known and it is not possible to back calculate the concentrations of the parent compounds. However, if cannabinol is not present and the chromatographic profile of cannabinoids is thought to be the same, then it is possible that the two or more cannabis samples being compared are related. This being the case, quantitative instrumental analysis would be required to confirm or otherwise the relationship of the samples. If, however, the thin layer chromatographic profile of the two samples is different, then an exclusion can easily be called and the cannabis samples may be confirmed to be unrelated. A further caveat to this is that the analyst should also consider whether or not the cannabis samples fit together. If two cannabis samples that are clearly related and fit together have been stored under different conditions, then it may well be that their cannabinoid profiles are different. This being the case, the relationship between the two samples can be established using a physical fit.

3.2.4.3 High-performance Liquid Chromatography of Cannabis Samples

The cannabinoids in cannabis samples can be separated, identified and quantified using HPLC. A number of systems have been described but care should be taken in their use because of the chemical lability of the cannabinoids – cannabinoids are labile in the presence of mineral acids. The vast majority of systems described use a buffered mobile phase in order to improve the quality of the chromatography. The reason for this is that compounds like Δ^9-THC and Δ^9-THC-COOH are weakly acidic as they contain phenolic groups and carboxylic acid groups that dissociate in the aqueous solution of the mobile phase. Dissociation (Figure 3.5) can be prevented by using a buffer containing a weak organic acid, for example, acetic acid.

A number of authors have used methanol, chloroform or methanol/chloroform mixtures to extract the cannabinoids from the material being analysed. These are sufficiently polar to extract the majority of the cannabinoids, but it is well known that cannabinoids are not stable in chloroform as exemplified in a recent study of Nyaope – a drug containing both heroin and cannabis.[3]

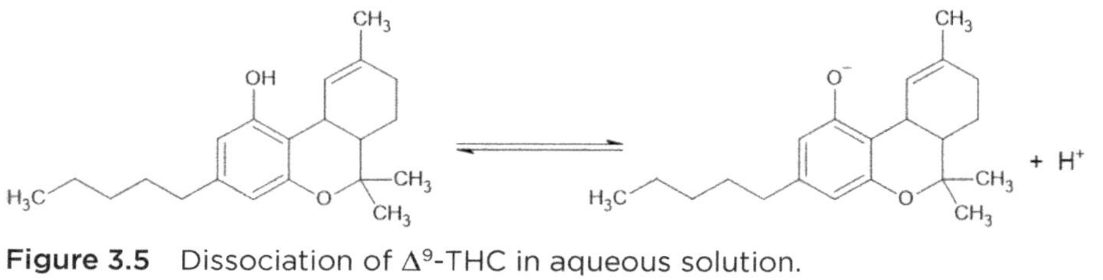

Figure 3.5 Dissociation of Δ^9-THC in aqueous solution.

A further requirement of the extracting solvent is that it is compatible with the mobile phase of the HPLC system. It is for these reasons that the methods recommended by the UNODC[8] are widely adopted and ethanol is adopted as the extracting solvent, although the UNODC do indicate that the solvent chosen may depend upon the analysis being carried out.[8]

One of the simplest methods is an ion-suppressed reversed-phase method, using a reversed-phase C-18 (ODS-2), 25 cm × 4.6 mm i.d., 5 µm particle size column, with a mobile phase of methanol/water/acetic acid 85/14.2/0.8 by volume, at a flow rate of 1.5 ml min^{-1} and detection at 230 nm. This has been used to successfully identify and compare cannabis samples.[6] Cannabis material was extracted in ethanol, at 10 mg ml^{-1} for 30 min at room temperature, the extract centrifuged at 5000g for 5 min to remove any particulate material and the supernatant subjected to the HPLC analysis. The extraction at 10 mg cannabis material per ml solvent was chosen so that, assuming that the material being analysed contained 10% Δ^9-THC, the extract would contain this drug at 1 mg ml^{-1}, in the same concentration range as the standards that can be commercially purchased, and a concentration that is easily analysed by HPLC. It is important to remove any particulate material to prevent blockage of the HPLC system. This system resulted in retention times of 4.16 min for cannabidiol, 6.49 min for cannabinol and 7.65 min for Δ^9-THC.

HPLC suffers from a number of drawbacks in terms of the analysis of cannabis. Due to the complex nature of cannabis samples, it is not always possible to achieve baseline resolution of every compound in the mixture being analysed. More often than not, the phenolic material will precipitate out of the mobile phase at the top of the guard column (which can be changed) or analytical column (which is expensive to replace), causing a blockage and poor chromatography. This is most frequently identified through a gradual increase in the back pressure in the HPLC system. Analysis times can be in excess of 1 h making sample throughput slow. Finally, there are large volumes of waste mobile phase to be disposed of. It is for these reasons that cannabis samples are analysed by GC–MS, with or without derivatisation.

3.2.4.4 Gas Chromatography–Mass Spectrometry of Cannabis Samples

Some laboratories use GC coupled to a flame ionisation detector (GC–FID) at 300 °C but such detectors while providing a chromatographic profile will not allow definitive identification of the drugs in question because only retention time data are available. The high resolution of analytes offered by gas chromatography coupled to the definitive identification offered by mass spectrometry makes GC–MS an ideal technique for the analysis and identification of cannabis. The method also requires thermal stability of the analytes and this condition is not met for all cannabinoids. For example, Δ^9-THC-COOH thermally decarboxylates to Δ^9-THC. This is not necessarily a challenge to the forensic analyst if total Δ^9-THC is to be measured. However, if the content of both Δ^9-THC and Δ^9-THC-COOH is required, then derivatisation is necessary.

An exemplar GC method is provided by the UNODC.[8] A 15 m × 0.25 mm i.d × 0.25 μm BP-5 or DB-5 column can be used, with hydrogen as the carrier gas at 1.1 ml min^{-1}. Some laboratories use nitrogen instead of hydrogen. The injection may be split or split less depending upon sample concentration, with an injection port temperature of 280 °C. This temperature is sufficient to volatilise all of the compounds of interest. A column oven temperature programme starting at 200 °C (2 min), rising from 10 °C min^{-1} to 240 °C and holding at 240 °C (2 min) can be employed. Most laboratories use variations on these conditions once their method has been validated.

Typical MS conditions include an interface temperature higher than the end of the oven temperature programme (to prevent sample condensation and retain the analytes in the vapour phase), a voltage energy of 70 eV and source and detector temperatures appropriate to the instrument being used. The scan rate should be sufficient to allow detection of all of the analytes of interest, and sufficiently high that if quantitative work is being undertaken that at least 10 scans are taken across a peak in order to ensure accurate determinations of the drugs in the samples. The cannabinoids in the sample are identified on the basis of their retention times (or retention index against an internal standard) and their mass spectra. Exemplar mass spectroscopic data are provided in Table 3.1.

GC–MS is a powerful tool for the identification of cannabinoids, for example, the isomers of THC. Although their retention times are close, they are easily differentiated on the basis of their mass spectra. For example, the abundance of the ion at m/z = 299 is

Table 3.1 GC–MS mass spectroscopic data illustrating how mass spectrometry can be used to differentiate between cannabinoids and isomers of THC. Data from ref. 8.

Cannabinoid	Ion (relative abundance) (%)
CBD	314 (7), 246 (13), 231 (100)
Δ^8-THC	314 (81), 299 (9), 271 (37), 258 (38), 231 (100)
Δ^9-THC	314 (85), 299 (100), 271 (46), 258 (23), 231 (70)
CBN	310 (11), 295 (100), 251 (4)

only 9% in the mass spectrum of Δ^8-THC, while the ion is the base peak in the mass spectrum of Δ^9-THC. Small differences in the absolute abundance of the ions in the mass spectra should be expected. This is due to a number of factors including being from different samples, with the instruments subject to different operating, maintenance and cleaning regimes. It is for this reason that positive controls should be run as well as the sample(s) to obtain mass spectra under as near identical conditions as possible and achieve definitive identification of the samples.

During the GC process, some cannabinoids, for example THCA, will thermally decarboxylate to Δ^9-THC. This is detected as Δ^9-THC and provides a measure of the total Δ^9-THC content. However occasionally, it is necessary to determine both compounds, for example, when samples are being profiled or compared. Under such circumstances, it is necessary to derivatise the samples. A number of reagents are available to derivatise cannabinoids. One such reagent is N,O-bis(trimethylsilyl)acetamide (N,O-BSA). In order to use this reagent an extract, or solution of standards, is prepared as described above in ethanol. An aliquot (200 µl is a convenient size) is dried under a stream of nitrogen. Nitrogen is used because it is chemically inert, oxygen and water free. To this is added the derivatisation reagent (100 µl of 10% N,O-BSA in n-hexane containing 0.1 mg ml^{-1} n-alkane internal standard). The sample vial is closed, the sample is shaken and allowed to derivatise at room temperature for up to an hour and subsequently analysed by GC–MS. Positive controls should be prepared for comparative purposes as should solvent and reagent blanks to demonstrate that the reagents that are used and the instrument are not contaminated with drugs. With the formation of trimethylsilyl ethers, the mass spectroscopic data are very different to those obtained from the native drugs, as shown in the exemplar data in Table 3.2. It is still possible, however, to use the mass spectra of the trimethylsilyl (TMS) derivatives of the isomers of THC to differentiate between the two compounds.

Table 3.2 GC–MS mass spectroscopic data for the native drugs and TMS derivatives of THC isomers. Data from ref. 8.

Cannabinoid	Ion (relative abundance) (%)
Δ^8-THC	314 (81), 299 (9), 271 (37), 258 (38), 231 (100)
Δ^8-THC – TMS	386 (69), 371 (8), 343 (22), 330 (44), 315 (3), 303 (100)
Δ^9-THC	314 (85), 299 (100), 271 (46) 258 (23), 231 (70)
Δ^9-THC – TMS	386 (98), 371 (100), 343 (29), 330 (16), 315 (53), 303 (45)

Other reagents are available that react with the phenolic and carboxylic acid groups on cannabinoids. These include *N*-methyl-*N*-trimethylsilyltrifluoroacetamide (MSTFA), *N,O*-bis(trimethylsilyl)trifluoroacetamide/trimethylchlorosilane (1%) (BSTFA/TMCS) and acid anhydrides. In each case, derivatives of standards, samples, reagent blanks and negative controls should all be prepared in order to demonstrate that any drug identified in the sample comes from the sample and nowhere else in the materials being analysed.

Cannabis samples can also be quantified. Calibration data can be obtained using different amounts of standard in an internal standard solution and subjected to GC–MS, or by derivatisation of different amounts of standard with a constant amount of internal standard added. The sample should then be subjected to the same preparative process, as should the relevant blanks and reagent controls and the data generated by GC–MS and subsequently analysed. Cannabis comparison can be achieved by GC–MS in the same way. Samples should be prepared in an identical manner and subject to GC–MS analysis. The chromatographic data will then be directly comparable.

3.2.4.5 The Analysis of Stable Isotopes to Determine the Origin of Cannabis

It has long been suggested that the sources of cannabis could be identified and compared by examination of stable isotopes.[9] Of particular interest are the stable isotopes of carbon (^{12}C and ^{13}C) and nitrogen (^{14}N and ^{15}N). The differences in the ratios of the stable isotopes ($^{13}C/^{12}C$ and $^{15}N/^{14}N$) are caused by the growing conditions of the cannabis.[9] For example, differences in ratios of carbon isotopes are greater in well-watered plants than they are in plants grown in dry regions. This is because of the relationship between the ratios in the atmosphere relative to those in the stomatal space in leaves, which is described by eqn (3.1).

$$\delta^{13}Cc_3 = \delta^{13}C_{atm} - a - (b - a) \times c_i/c_a, \qquad (3.1)$$

where $\delta^{13}Cc_3$ is the carbon isotopic composition of the atmospheric carbon dioxide, a is the physical fractionation of the carbon dioxide from the air to the stomatal chamber and b is the fractionation involved in the photosynthetic pathway of the plant – in the case of *Cannabis sativa,* this is the Calvin cycle.[9] The ratio of c_i/c_a is controlled by photosynthetic activity – a smaller ratio will cause a higher ratio difference. Differences in nitrogen ratios can be affected by different fertiliser regimes.[9] Such differences have been applied to determining the country of origin, region of origin and growing conditions where the cannabis was grown.

The methods can be applied to all parts of the cannabis plants.[9] It has been demonstrated that different parts of the plants including the leaves, seeds and twigs all contain the same isotope ratios when the source was the same plant. The material to be examined is washed in deionised water and dried at 40 °C for 24 h. The effect of the elevated temperature on cannabinoids does not matter because it is not the cannabinoids that are being examined. The plant material should be ground to a fine homogeneous powder that represents the plant material as a whole. Around 10 mg of plant material is required for analysis. In real terms, this is quite a large sample and represents one of the disadvantages of this method when the material is scarce. Once washed and dried, the cannabis material is analysed using continuous flow isotopic ratio mass spectrometry. The organic matter is combusted generating nitrogen and carbon dioxide. These are separated chromatographically in highly pure helium and analysed in an isotope ratio mass spectrometer. The nitrogen and carbon isotope ratios are measured and the differences expressed as δ (parts per thousand) using eqn (3.2).

$$\delta = ((R_{sample}/R_{standard}) - 1) \times 1000 \qquad (3.2)$$

where R_{sample} is the isotopic ratio of the samples and $R_{standard}$ is the isotopic ratio of the standards gas. The ratios are confirmed by comparison with calibrant gas ratios. It is then possible to apply, for example, clustering techniques for sample comparison. A number of studies have been undertaken to demonstrate that origin of cannabis can be identified using stable isotopes. For example, cannabis growing in Australia, Papua New Guinea and Thailand has been examined.[10] The $\delta^{13}C$ values ranged from −36 to −25‰ and the $\delta^{15}N$ values ranged from −12 to +15.8‰. It was not possible to determine the country of origin of the cannabis but it was possible to determine whether the cannabis had been grown under well-irrigated conditions or in a drier environment. Similarly, the differences in nitrogen ratios were affected by the fertiliser regimes under which the cannabis was grown. This meant that using the differences in nitrogen ratios, it may be possible to relate cannabis samples to a particular crop.

It has been shown that the plant material contains the same isotope ratios regardless of the part of the plant being examined.[9] It was also possible to identify those plants that have been grown in dry regions and in wetter regions. However, when dry regions were compared with each other, and also wet regions with each other, there was some overlap between different dry regions and different wet regions in terms of their stable isotopic ratios. On this basis, it is not currently possible, with court-going accuracy and precision, to identify the source of cannabis based on stable isotopes alone. Where this technique is perhaps more useful is in confirming that cannabis plants may have originated from a particular geographic location for intelligence purposes and achieving an exclusion of a particular geographic location as a source of the plants.

3.2.4.6 The Use of DNA to Identify Cannabis

It has been shown that it is possible to identify cannabis using DNA techniques,[11] although this is not routinely carried out in operational laboratories. It is useful, however, to understand the principles should a chemical technique not be found to identify a plant-based drug for court-going purposes. By way of example, in order to achieve identification of *Cannabis,* the variation in DNA sequence in the intergenic spacers between two highly conserved genes in the chloroplasts can be examined. Using this methodology, a *Cannabis*-specific sequence was found between the two chloroplast tRNA genes, trnL and trnF, consisting of the 6 base pair repeat sequence 5'-CACCAT-3'. This was found to be entirely *Cannabis* specific and occur with between 3 and 40 repeats. No cross reaction from any other plant species occurred. This allows forensic scientists to extract the chloroplast DNA from *Cannabis* materials, amplify it using the polymerase chain reaction (PCR) and subject it to DNA analysis. If amplification occurs, then the plant material is definitely *Cannabis*, and if the same number of repeats of the *Cannabis*-specific sequence is observed, the plant material may well be related. This technique is useful in a number of different ways. It is, first, possible to identify *Cannabis* when there are no macroscopic or microscopic features available to make a formal botanical identification. Similarly, for plant material, it is possible to identify the material as *Cannabis* where chemical markers are missing through, for example, oxidation or other decomposition. Third, it is possible to determine or otherwise the relatedness of *Cannabis* samples when the number of repeats of the *Cannabis*-specific sequence is determined.

References

1. Available from: https://wdr.unodc.org/wdr2020/field/WDR20_Booklet_3.pdf. Accessed 11 Jul 2022.
2. Available from: https://www.emcdda.europa.eu/media-library/drug-seizures-glance-2019_en. Accessed 11 Jul 2022.
3. P. M. Mthembi, E. M. Mwenesongole and M. D. Cole, *S. Afr. J. Sci.*, 2021, **117**(11–12), 1.
4. T. A. Gough, in *The analysis of drugs of abuse*, ed. T. A. Gough, John Wiley and Sons, Chichester, 1991, pp. 540–541.
5. B. E. Buchanan and D. O'Connell, *Sci. Justice*, 1998, **38**, 221.
6. R. Gillan, M. D. Cole, A. Linacre, J. W. Thorpe and N. D. Watson, *Sci. Justice*, 1995, **35**, 169.
7. Available from: https://www.unodc.org/unodc/en/data-and-analysis/bulletin/bulletin_1978-01-01_2_page008.html. Accessed 18 Jul 2022.
8. Available from: https://www.unodc.org/documents/scientific/Recommended_methods_for_the_Identification_and_Analysis_of_Cannabis_and_Cannabis_products.pdf. Accessed 18 Aug 2024.
9. E. K. Shibuya, J. E. S. Sarkis, O. N. Neto, M. Z. Moreira and R. L. Victoria, *Forensic Sci. Int.*, 2006, **160**, 35.
10. T. M. Denton, S. Schmidt, C. Critchley and G. R. Stewart, *Aust. J. Plant Physiol.*, **28**(10), 1005.
11. A. Linacre and J. Thorpe, *Forensic Sci. Int.*, 1998, **91**(1), 71.

4 The Analysis of Synthetic Cannabinoids

4.1 Introduction

Synthetic cannabinoids, otherwise known as synthetic cannabinoid receptor agonists (SCRAs), first appeared in the illicit drug market around 2008. Between 2008 and 2020, the EMCDDA reported that 209 compounds have been described in the forensic science literature. The rate of SCRA appearance peaked at around 27 new compounds a year, and currently around 10 or 11 new compounds are described annually. The first report of the synthesis of analogues of Δ^9-tetrahydrocannabinol (Δ^9-THC) was in 1978, and subsequently a number of analogues, which bind to cannabinoid receptors in humans, were prepared as experimental drugs for the alleviation of conditions including pain, anorexia, muscle spasms and glaucoma. However, these drugs were found to bind to cannabinoid receptors more strongly than the drugs in cannabis with the additional problems associated with their psychotropic effects. The effects, similar to those caused by the drugs in cannabis, are much stronger and last between 1 and 6 h. The onset time for such drugs depends on the drug in question, the route of administration and the amount consumed. For example, drugs marked as "Spice" start to have an effect in seconds, peak in about half an hour and the effects then tail over 2 to 3 h.

The first *N*-alkyl indole-3-carbonyl derivative was JWH-018, which followed the appearance of HU-210, a synthetic example of a classic cannabinoid and cyclohexylphenols (Figure 4.1). Following the appearance of this drug, there have been a large number of synthetic

Controlled Drug Analysis
By Michael D. Cole, Lata Gautam and Agatha Grela
© Michael D. Cole, Lata Gautam and Agatha Grela 2025
Published by the Royal Society of Chemistry, www.rsc.org

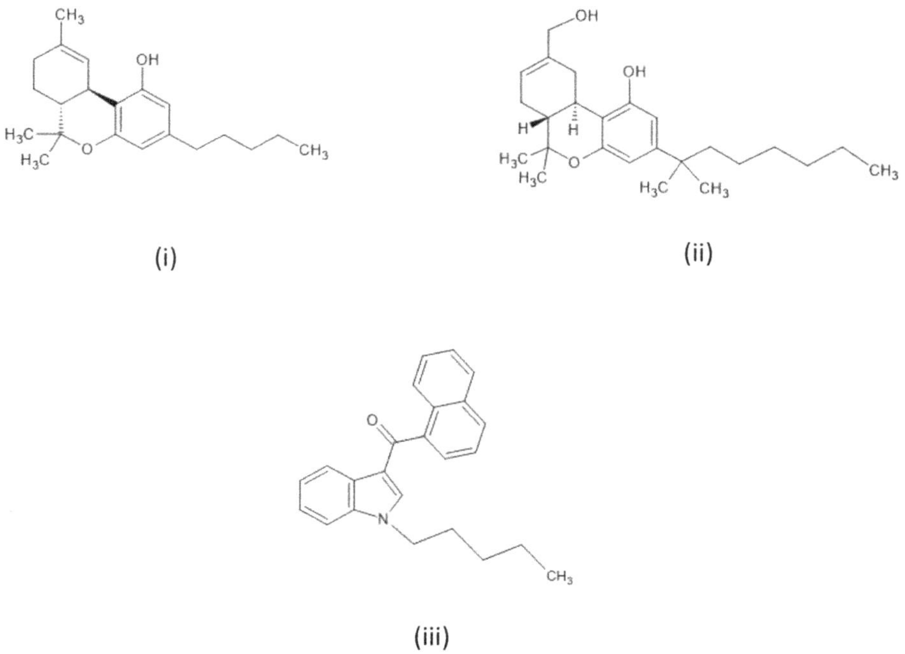

Figure 4.1 Structures of (i) Δ^9-tetrahydrocannabinol, (ii) HU-210 and (iii) JWH-018.

cannabinoids manufactured and distributed, often marketed over the internet as "herbal incense". These drugs are cannabinomimetics, mimicking the pharmacological action of cannabinoids, but they are much more potent in their action.

Chemically, synthetic cannabinoids are a very diverse group of compounds (Figures 4.1 and 4.2). There are no current data on the actual methods used for the clandestine manufacture of these drugs, although there are a large number of syntheses reported in the refereed literature. The cyclohexylphenols, of which CP47,497 is an example, are similar to Δ^9-THC in the C ring and they possess the alkyl chain, as on the C ring of the cannabinoids, which is important for interaction of the drugs with the chemical receptors for these drugs on the surfaces of cell membranes. The naphthoylindoles (*e.g.* JWH-018), also known as alkylamino indoles, contain naphthyl and indole rings joined by a carbonyl bridge. They are *N*-alkylated with chains of varying lengths. The naphthylmethylindoles (*e.g.* JWH-175) are similar in structure to the naphthoylindoles but the carbonyl group is absent. Similarly, the

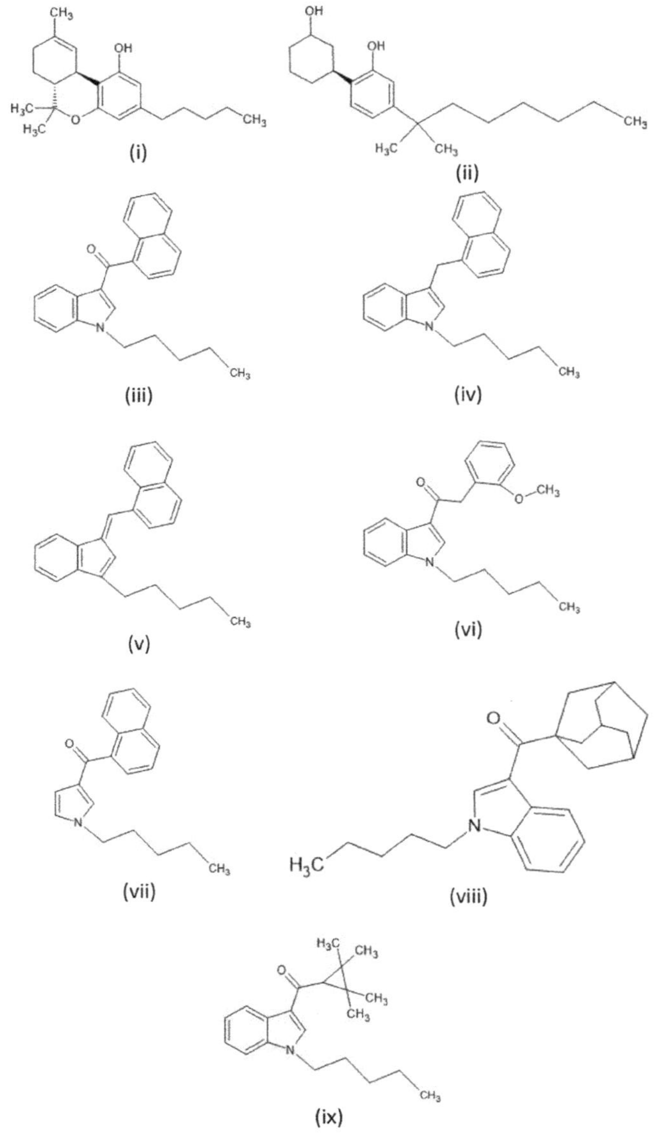

Figure 4.2 Structures of (i) Δ⁹-tetrahydrocannabinol, (ii) CP47,497, (iii) JWH-018, (iv) JWH-175, (v) JWH-176, (vi) JWH-250, (vii) JWH-30, (viii) AB-001 and (ix) UR-144.

naphthylmethylindenes (*e.g.* JWH-176) are structurally similar to the previous two classes of drugs, but the indole ring is replaced by an indene ring, with a double bond in the chemical bridge, with *E* orientation of the substituents. The benzoylindoles, exemplified by

JWH-250 have substitutions on both the indole ring and phenyl ring. Another group, the naphthoylpyrroles (*e.g.* JWH-030) are similar to the naphthoylindoles but the indole group in the latter is replaced by a naphthoyl substituent. The phenylacetylindoles (*e.g.* JWH-250) are similar to the benzoylindoles but the bridge between the two ring systems is extended by the presence of a methylene group. A further group of drugs, considered "third-generation drugs" in this group, appeared from 2012 – the adamantoylindenes, as exemplified by AB-001. These contain an adamantyl ring that replaces the naphthyl system present in previous examples. A further group of drugs, the tetramethylcyclopropylindoles (*e.g.* UR-144), have emerged. Going forward, it is likely that other chemical groups will be developed.

The drugs are sold widely over the internet. The drugs are prepared as solutions and then sprayed onto plant materials, which are then dried, leaving the drug residue on the surface of the plant material. The plant materials are then mixed with tobacco, and smoked. These materials are often perfumed and sold marked "not for human consumption". Common plant substrates include Damiana (*Turnera diffusa*) and Marsh Mallow (*Althaea officinalis*). The mixture, dried and packaged with 1–3 g of the plant material sprayed with a solution of the drug, is then packaged in plastic or mylar (a form of polyester) bags (Figure 4.3). Due to the way that the materials are prepared, they are very heterogeneous – the covering of the drug on the plant material is not at all even. On purchase by the users, these materials are frequently mixed with tobacco and smoked. Typical amounts of drug on the plant material range from 5–20 mg g^{-1}, but there is much variation in the drug content on different substrates. Coefficients of variation of drug concentration can be up to 20%. This needs to be taken into account when preparing materials for analysis.[1] One way to address this problem is to homogenise the whole sample so that a representative value for drug concentration may be achieved, but this is at the loss of the physical integrity of the sample and any morphological information that the sample might provide. Recent reports of newer dose forms and ways to smuggle the drugs, for example into prisons, include paper and clothing impregnated with synthetic cannabinoids.[2] Some materials may contain one of the drugs, others may contain mixtures and some with the same name may contain only one drug in one batch but multiple drugs in others.

Control of this group of compounds is complex. Many are controlled in the United States as Schedule I drugs, while in the United

 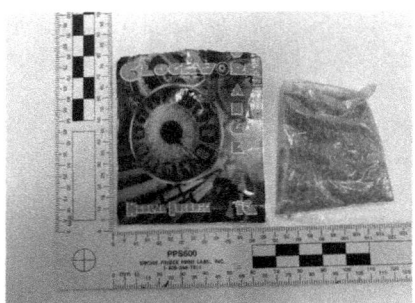
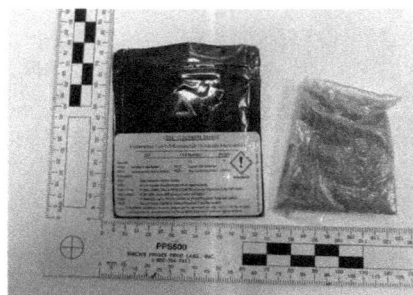 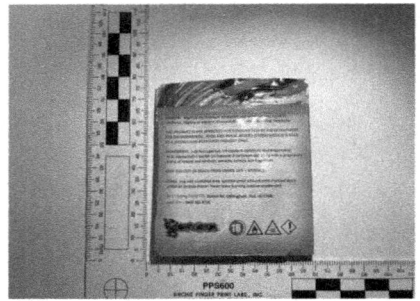

Figure 4.3 Examples of packaging and herbal drugs containing synthetic cannabinoids.

Kingdom, many are Class B drugs as defined by the Misuse of Drugs Act 1971.

Based on the above, chemical analysis is always required to identify the drug present in the sample – reliance cannot be placed on the name of the drug alone or on the identification of the plant material onto which the drug has been sprayed.[1]

The analysis of synthetic cannabinoids is rendered complicated by the structural variety of the drugs that may be encountered. Analysis includes a physical description of the drug material as seized which includes a description of the drug material itself, the packaging and whether or not the packaging of multiple items might be related. Where this is possible, it may be feasible to relate the packaging of different samples to each other or to other cases. Subsequent analysis then involves colour tests, thin layer chromatography, and instrumental methods including HPLC and/or GC–MS. Where new classes of these drugs are encountered, it may also be necessary to apply one and two-dimensional NMR techniques to determine the structure(s) of the drugs involved.

4.2 Analysis of Synthetic Cannabinoids

4.2.1 Physical Description

Prior to the analysis, it is essential that the sample is subjected to a full physical description and prepared for subsequent analysis. Both the drug and its packaging should be described. For the packaging, this should include size, nature and material of the packaging itself, any logos or pictures on the packaging and any text including where possible a description of the font and font size. This may allow relationships to be established with materials from other cases. The plant material contained within the packaging should also be described including details of, for example, weight and any botanical features that might be present.

Once the material to be analysed has been chosen, a sample of it should be homogenised. This can be achieved using a pestle and mortar, a powder mill or, as in a number of studies, ground between two pieces of sandpaper.[3] In all cases, the relevant precautions and controls should be undertaken to determine and demonstrate that any synthetic cannabinoid identified comes from the drug sample and from no other source. It is also important that drugs are analysed as soon as possible after being seized. Synthetic cannabinoids containing ester, amide and constrained ring systems will all undergo hydrolysis in the presence of atmospheric water and it is theoretically possible that they will oxidise too in the presence of atmospheric oxygen. Under such circumstances, the compounds present at the point of analysis may not be the same as those at the point of seizure, much as for cannabis materials themselves. Once the physical description of the sample has been undertaken, the next step in drug identification is chemical analysis.

4.2.2 Presumptive Tests for Synthetic Cannabinoids

Due to the complex nature of this group of drugs, there is no single presumptive (colour) test for this group of drugs that can be universally applied to identify them. This is because there is no single chemical functional group or group of functional groups on these molecules that allows their identification using colour tests. The traditional colour test for natural cannabinoids, the Duquenois–Levine test, does not hold good for the synthetic cannabinoids. This is because the Duquenois–Levine test is a

Table 4.1 Examples of presumptive tests for different functional groups found within synthetic cannabinoid receptor agonists that give positive reactions.

Presumptive test	Functional/ chemical group	Examples of SCRAs that the test reacts with
2,4-Dinitrophneylhydrazine (2,4-DNP)	Carbonyl group	Naphthoylindoles
Marquis reagent	Some classes of alkaloid	Phenylacetylindoles Cyclohexylphenols JWH compounds
Fast blue BB	Phenols	Cyclohexylphenols

test for phenolic groups, which are not universal in this type of drug. Similarly, the van Urk reagent, which is used for the detection of indole-containing drugs, for example, lysergic acid diethylamide (LSD), does not give a positive colour reaction with this drug group. Again, this is due to the fact that these drugs do not contain the necessary chemical moieties to react with this reagent. However, colour tests do provide the opportunity to tentatively *exclude* a particular type of synthetic cannabinoid if a chemical reaction does not occur. The exclusion is only tentative because colour tests are relatively insensitive and the samples are heterogeneous. The reactions of the various classes of synthetic cannabinoids with common presumptive tests for drugs of abuse are summarised in Table 4.1. The conclusions drawn from these tests should always be tested using thin layer chromatography and/or confirmed using instrumental methods where standards are available.

4.2.3 Thin Layer Chromatography of Synthetic Cannabinoids

Thin layer chromatography is another method that has been applied to the analysis of synthetic cannabinoids. Prior to analysis, a solution of the sample must be prepared. In general, the synthetic cannabinoids are soluble in the traditional low-molecular-weight organic solvents. However, these drugs are prone to hydrolysis of ester and amide groups. Similarly, the constrained ring systems, for example, the tetramethylcyclopropyl ring system found in UR-144, are subjected to chemical decomposition. Analysts should review their findings on identification of the drug to ensure that no artefact formation or drug decomposition could have occurred during the analysis by considering the structure of the drug and the chemistry of the analytical process throughout the whole analytical life cycle.

The most frequently used chromatographic systems are normal-phased systems on silica gel. The details of these systems are given in Table 4.2. Bases are added to the mobile phase, for example, diethylamine or small amounts of concentrated ammonium hydroxide, to achieve ion suppression, related to the nitrogen content of the drugs, so that the chromatography is improved and tailing reduced. Once the chromatogram is developed and analytes separated, the drugs are visualised in a number of ways. Detection under UV light can be achieved because the conjugated ring systems on the drugs will strongly absorb UV light at 254 nm. Such compounds will appear as purple spots against a coloured background on thin layer chromatography plates impregnated with a fluorescent compound. Sulfuric acid and nitric acid can be used as spray reagents to dehydrate the compounds giving yellow/brown dehydrated products against a pale yellow or white background. However, these acids will react/dehydrate with most compounds and are very non-specific. Fluorescamine reacts with primary amino groups forming a fluorescent pyrrolinone product that can be detected under UV light at 366 nm. Ninhydrin reacts with primary amines to give a purple product (Ruhemann's purple), secondary amines to give a yellow product and with some tertiary amines to give a grey-blue product.

Table 4.2 Exemplar TLC systems that can be used for the analysis of synthetic cannabinoid receptor agonists. The combination of mobile phase and visualisation methods will depend on the SCRAs being analysed.[4]

Stationary phase	Mobile phase (solvent composition by volume)	Visualisation reagents
Silica gel	Toluene : diethylamine 9 : 1	UV (254 nm)
	Ethyl acetate : dichloromethane : methanol: conc. NH_4OH 18.5 : 18 : 3 : 1	UV (366 nm)
		Fluorescamine (50 mg in 1 L acetone); spray, then view at 366 nm
		10% sulfuric acid; spray, then view at 366 nm
		Potassium iodoplatinate (5 g chloroplatinic acid and 35 g potassium iodide in 1650 ml water and 49.5 ml conc. HCl); spray and view under white light

Dianisidine will form azo dyes with some aromatic compounds. Potassium iodoplatinate reacts with tertiary and quaternary amino compounds. These reagents react with part of the drug molecule but the functional groups or part of the molecule that they react with occur very widely and as a consequence, none of these tests will be definitive and those with which the reagents will react with need to be determined experimentally, usually spectroscopically.

Another challenge facing the forensic scientist when analysing synthetic cannabinoids by TLC is the lack of resolution of the compounds in this diverse class of drugs. It is known that the R_f value of a drug under TLC analytical conditions is affected by, for example, small changes in composition of the mobile phase (which in turn affects the polarity and pH of the mobile phase), the degree of saturation of the atmosphere of the chromatography tank and the temperature of the chromatographic system – as temperature increases so does the R_f value of the analyte. For example, JWH-081, JWH-251 and JWH-398 (Figure 4.4) are reported to have the same R_f value in the first of the TLC systems described and their absorption of UV light and reaction with the visualisation reagents are very similar.[1] On this basis, synthetic cannabinoids cannot always be identified using TLC.

4.2.4 Gas Chromatography of Synthetic Cannabinoids

Synthetic cannabinoids can be analysed using standard GC–MS systems (Table 4.3). In such systems, the synthetic cannabinoids are analysed as either native drugs or following derivatisation. Derivatisation reagents that have been used include *N,O*-bis(trimethylsilyl)trifluoroacetamide/trimethylchlorosilane (BSTFA/TMCS).[1] Acid anhydrides can be conveniently used because they do not result in the complex mixture of reaction by-products associated with, for example, *N,O*-bistrimethylsilylacetamide. In such systems, the drugs or their derivatives are identified on the basis of their retention time or retention index relative to an internal standard, when compared to known compounds and comparison of mass spectra between sample and standard, or with databases.

Most drugs are identifiable by GC–MS with unique retention times and mass spectra. There are some exceptions – for example, the isomeric pairs HU-210 and HU-211 and WIN 55,212-2 and WIN 55,212-3 (Figure 4.5). Unfortunately, derivatisation does not solve the problem for the HU-210 and HU-211 pair, where the mass spectrum of the di-TMS derivative remains near identical.

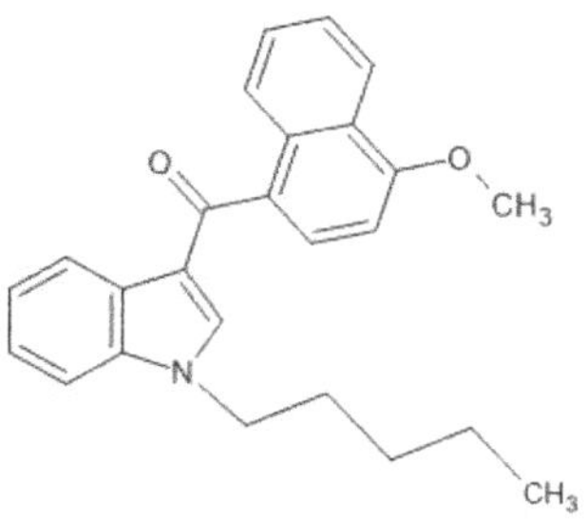

Figure 4.4 Structure of (i) JWH-081, (ii) JWH-251 and (iii) JWH-398.

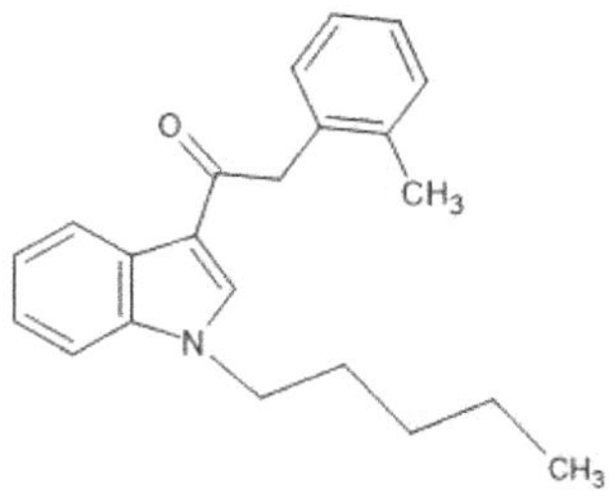

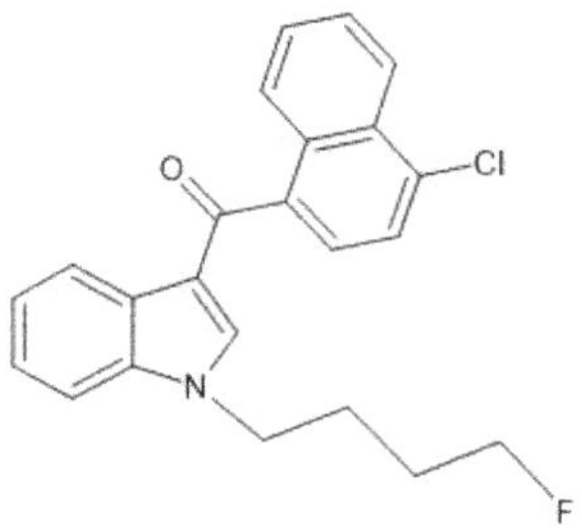

Table 4.3 Gas chromatographic systems that can be used for the separation and identification of synthetic cannabinoid receptor agonists.[1] Slight variation in the parameter might be required depending on the details of the instrument used and the drugs being analysed.

Parameter	Underivatised samples	Derivatised samples
Column	DB-1 or equivalent	DB-1 or equivalent
	12 m × 0.2 mm × 0.33 µm	12 m × 0.2 mm × 0.33 µm
Injection volume	1 µl	1 µl
Split mode	Splitless	Splitless
Temperature programme	50 °C, rising at 30 °C min^{-1} to 340 °C (2 min)	80 °C, rising at 25 °C min^{-1} to 340 °C (2 min)
Inlet temperature	265 °C	265 °C
Transfer line temperature	300 °C	300 °C
Carrier gas	He	He

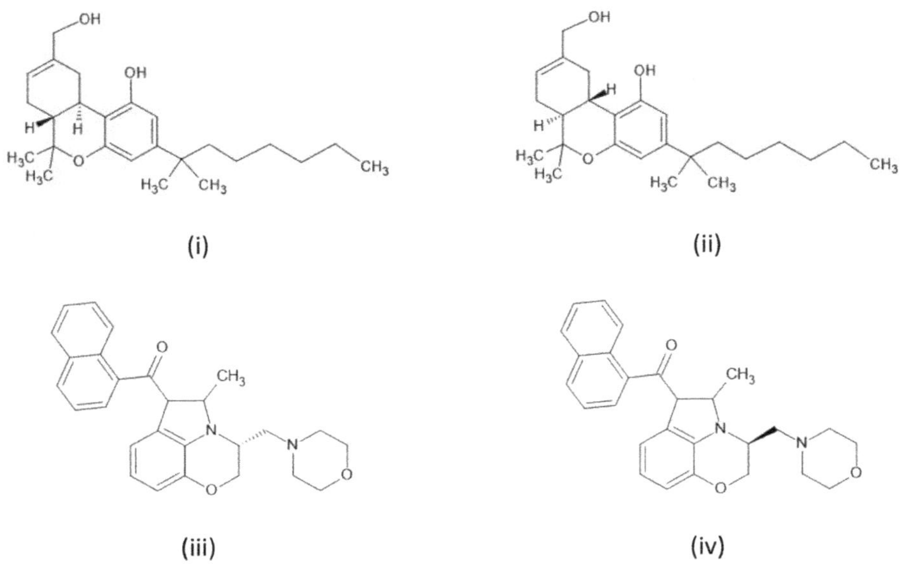

Figure 4.5 Structure of (i) HU-210, (ii) HU-211, (iii) WIN 55,212-2 and (iv) WIN 55,212-3.

It is also possible that the synthetic cannabinoids interconvert during analysis by GC–MS. For example, it is possible for AM-2201 to thermally decompose into JWH-018 and JWH-022 (Figure 4.6) under GC–MS conditions. Similarly, constrained ring systems such as the tetramethylcyclopropyl ring in UR-144 are known to open under GC–MS conditions.

Another problem with the GC–MS analysis of this group of drugs is that some will undergo transesterification. As an example, when dissolved in methanol or ethanol and subjected to GC–MS analysis, the heat of the injection block and analytical system causes the quinolinyl carboxylate PB-22 (an analogue of JWH-018) to undergo transesterification.[5] Forensic chemists should be aware of such problems and choose the solvent for sample preparation appropriately.

4.2.5 High-performance Liquid Chromatographic Methods

A number of HPLC methods have been described for the analysis of street samples of synthetic cannabinoids. The samples can be dissolved in the HPLC mobile phase or in methanol, ensuring that the drug remains in solution when injected into the HPLC system.

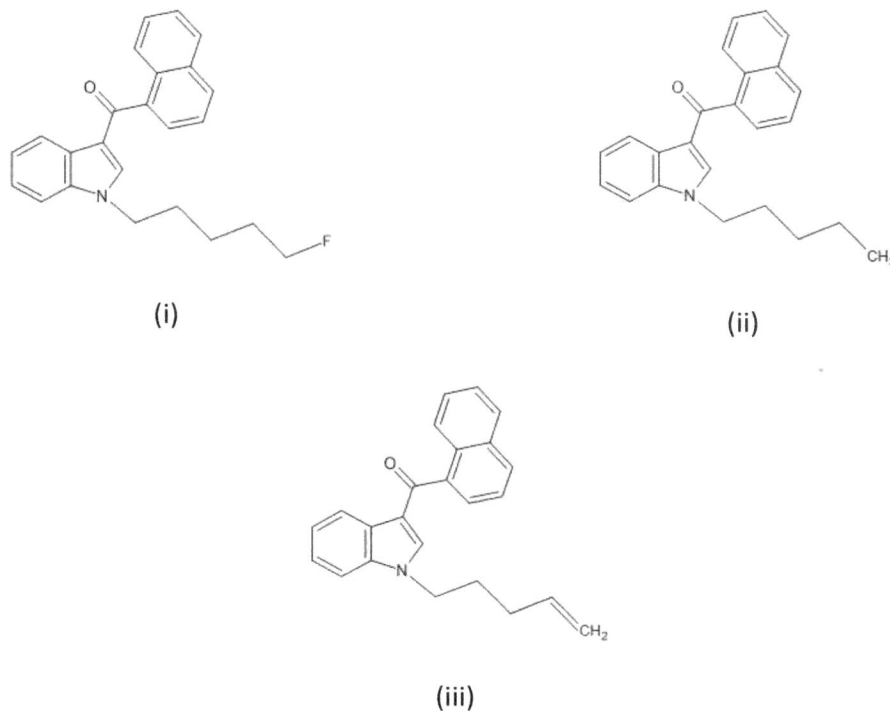

Figure 4.6 Structures of (i) AM-2201 and thermal decomposition products (ii) JWH-018 and (iii) JWH-022.

The HPLC systems described are all reversed-phase systems, with generally different buffers used to achieve an ion pairing separation mechanism. The mobile phase systems are either isocratic or involve a gradient programme depending on the complexity of the drug mixture being separated. It should be remembered that in street drug samples, the drug mixture is only likely to be composed of one, two or three synthetic cannabinoids and complex mobile phase gradient programming is not required.

Baseline or near-baseline resolution has been achieved for the analysis of synthetic cannabinoids in the studies described. However, HPLC offers further advantage over GC–MS because transacetylation and thermal degradations that have been described in methanol solutions of synthetic cannabinoids analysed using this latter method are not reported when HPLC is used. The separated compounds can also be recovered for subsequent structural analyses – for example, NMR. Indeed, preparative HPLC (a scaled-up version of analytical HPLC systems) has been used to recover sufficient drug from street samples when the structure or stereochemistry of the drug is unknown.

4.3 NMR Analysis of Synthetic Cannabinoids

Although GC–MS is widely used in drug identification and quantification, it has a number of drawbacks as described above. In addition, when new compounds appear in the market, it is unlikely that there will be reference data available. Under such circumstances, NMR is a technique that will find useful application. It provides structural information and quantitative data with the added advantage that the sample can be recovered at the end of the analysis. The technique also has the speed and sensitivity required in operational laboratories without the need for complex sample preparation and clean-up. Finally, databases of spectroscopic data can easily be constructed allowing comparison and exchange of data between laboratories.

NMR studies of synthetic cannabinoids have also been undertaken.[4,6,7] In these, each standard was analysed by high-field NMR (field strengths of upwards of 400 MHz), with the samples having been dissolved in deuterated solvents including $CDCl_3$. Street samples can be prepared by extraction into the same solvent, although manual shaking and mechanical vortexing may be required to assist in the dissolution process. Removal of any solid material, by centrifugation or filtration, will be required in order to ensure that particulates do not interfere with the quality of the spectrum obtained.

In a similar fashion, it is possible to quantify synthetic cannabinoids by NMR. In order to achieve accurate results, it is necessary that the relaxation delay between signal pulses is more than the relaxation time for the nucleus being measured. If this is not so, the nucleus will not relax fully and an underestimate of the drug concentration may be made. For indole-containing alkaloids, the required relaxation time was found to be 4 s,[4] but for accurate results this should be confirmed for each of the classes of synthetic cannabinoids being quantified. Maleic acid can be used as an internal standard and d_6-acetone the solvent, although care should be taken when using ketones as a solvent for drug classes containing primary amines where nucleophilic attack can occur. Signal integrals for calibration standards can then be used to obtain calibration data against which the sample data are compared.

References

1. B. K. Logan, L. E. Reinhold, A. Xu and F. X. Drummond, *J. Forensic Sci.*, 2012, **57**, 1168.
2. C. Norman, C. Walker, B. McKirdy, C. McDonald, D. Fletcher, O. B. Sutcliffe, N. NicDaeid and C. McKenzie, *Drug Test. Anal.*, 2020, **12**, 538.
3. A. Namera, M. Kawamura, A. Nakamoto, T. Saito and M. Nagao, *Forensic Toxicol.*, 2015, **33**, 175.
4. F. Fowler, B. Voyer, M. Marino, J. Fuizel, M. Veltri, N. M. Wachter and L. Huang, *Anal. Methods*, 2015, **7**, 7907.
5. J. N. A. Tettey, C. Crean, J. Rodrigues, T. W. A. Yap, J. L. W. Lim, H. Z. S. Lee and M. C. Ong, *Forensic Sci. Int. Synergy*, 2021, **3**, 100219.
6. S. J. Dunne and J. P. Rosengren-Holmberg, *Drug Test. Anal.*, 2016, **9**, 734.
7. M. A. Marino, B. Voyer, R. B. Cody, J. Dane, M. Veltri and L. Huang, *J. Forensic Sci.*, 2017, **61**, 582.

5 The Analysis of Hallucinogenic Drugs from Plants and Fungi

5.1 Introduction

Plants and fungi have been used for medicinal and religious purposes for millennia because of their hallucinogenic properties. Plants and fungi are a rich source of tryptamine- and indole-based drugs, including lysergic acid diethylamide (LSD), psilocin and psilocybin, N,N-dimethyltryptamine (DMT) and mescaline (Figure 5.1), all of which are strictly controlled (Table 5.1). In the first half of the 20th century, these drugs were investigated for their possible therapeutic properties, especially in psychiatry. Once the psychedelic compounds were successfully extracted from the plants, their properties became known and they became more widely available as they emerged on the drug scene and their use as recreational drugs increased. Perhaps the most well known of these is LSD. LSD is one of the most potent hallucinogens known, where only microgram amounts are required to cause a hallucinogenic effect in humans.

The drugs are found in a variety of dose forms. They can be incorporated into paper sheets, gelatin blocks, other "edibles", tablets and powders. As such, they present a number of interesting challenges to the drug analyst. In addition to unusual matrices, the chemical and analytical difficulties include thermal dephosphorylation of some of these drugs under GC–MS conditions and chemical shifts in NMR spectra being affected by the pH of the solution in the NMR tube. The solutions to such challenges are discussed below.

Controlled Drug Analysis
By Michael D. Cole, Lata Gautam and Agatha Grela
© Michael D. Cole, Lata Gautam and Agatha Grela 2025
Published by the Royal Society of Chemistry, www.rsc.org

The Analysis of Hallucinogenic Drugs from Plants and Fungi

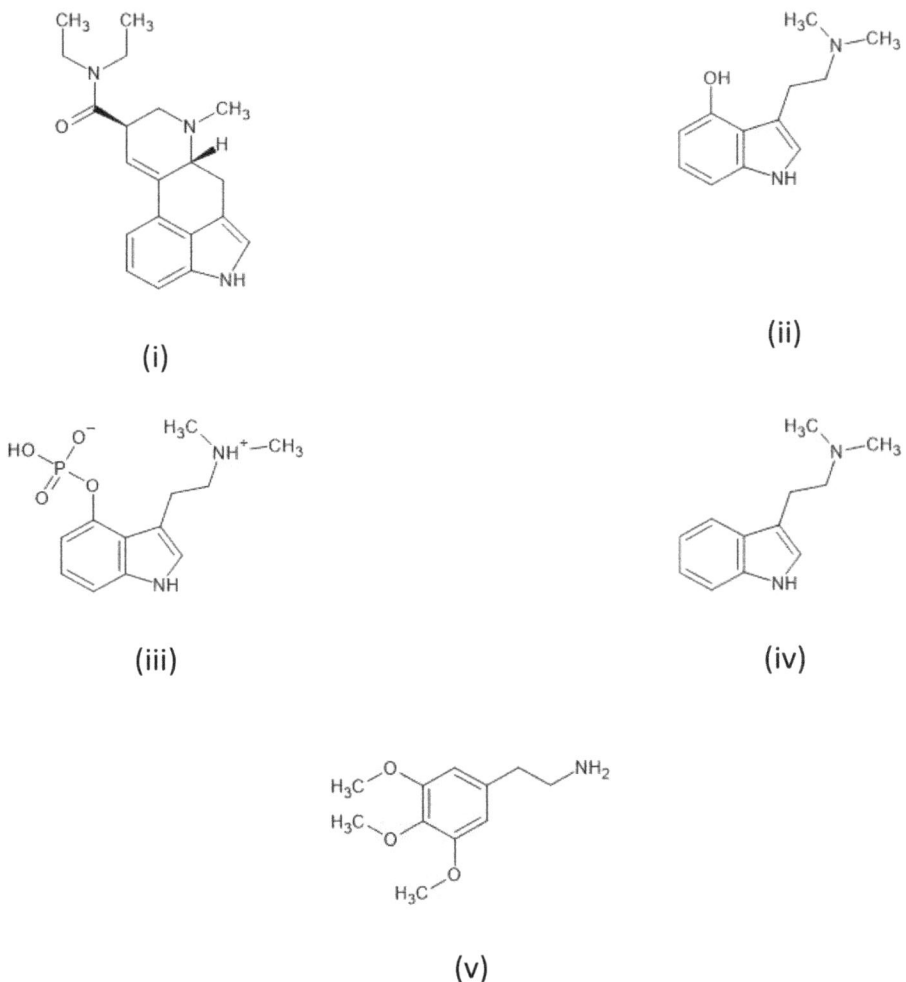

Figure 5.1 Structures of (i) LSD, (ii) psilocin, (iii) psilocybin, (iv) N,N-dimethyltryptamine and (v) mescaline.

5.2 The Analysis of LSD

LSD was discovered as a consequence of research into the medicinal properties of ergot fungi. LSD was first synthesised in 1938 and its hallucinogenic properties were discovered 5 years later. LSD is colourless, odourless, and water soluble. Originally encountered in gelatin blocks, which were swallowed by the drug users, it is now more commonly found soaked into sheets of absorbent paper that are dried later. These are commonly printed with distinctive designs and

Table 5.1 Legislative status of exemplar hallucinogens derived from plants and fungi.

Drug	Australia	United Kingdom	United States	New Zealand
LSD	Schedule 9	Class A	Schedule 1	Class A
Psilocybin	Schedule 8/9	Class A	Schedule 1	Class A
Psilocin	Schedule 9	Class A	Schedule 1	Class A
N,N-DMT	Schedule 9	Class A	Schedule 1	Class A
Mescaline	Schedule 9	Class A (plants containing mescaline are legal)	Schedule 1	Class A

perforated so they may be torn into single small squares each containing a single dose that are swallowed by the user. Paper is now preferred over gelatin because the homogeneity of the dose in each dose unit can be more easily controlled – which is important when the drug is so potent. The paper dose forms are commonly known as "blotters", "blotter acid" or "paper squares" (Figure 5.2). The drug may also be found in chocolate candy "kisses", where the drug can be dropped onto the surface of aspirin tablets as a solution with subsequent drying and as gelatine capsules. The drug solution is mixed with liquid gelatin which

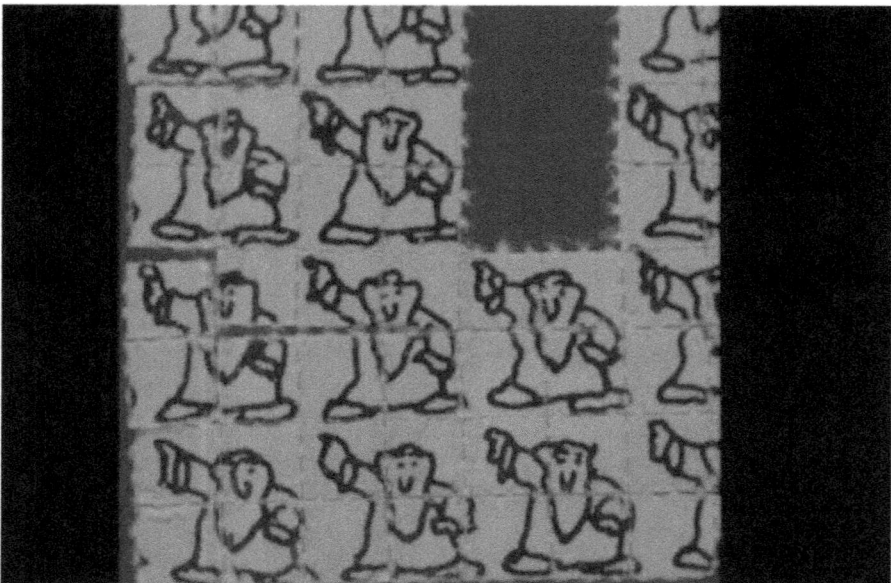

Figure 5.2 An example of a "blotter" acid into which a solution of LSD has been soaked and then dried.

is then cooled and allowed to set, or can be found in the form of pressed coloured scored tablets. Doses in each individual dose unit vary in range from 31 to 95 µg per unit and the effects are reported to start within 40 min and last up to 12 h.

While preparing the samples for analysis, and during the analysis itself, the drug chemist should be aware of a number of difficulties associated with the analysis of LSD. LSD degrades readily, particularly in biological specimens, unless protected from light and elevated temperatures. It may also bind to glass containers in acidic solutions, which would, for example, affect concentration measurements. The only analogue of LSD to have received widespread interest is the N-methylpropylamide of lysergic acid (LAMPA) (Figure 5.3), and any analytical technique should be capable of separating LAMPA from LSD. Analysts should also be aware that LSD, including powder, is "sticky" and tends to remain on gloves, clothing and the work bench. LSD also penetrates the skin very easily. Therefore, when handling LSD, especially in powdered form, personal protective equipment (PPE) most be worn. After the analysis is completed, the work area should be cleaned using diluted bleach and be examined under UV light, checking for fluorescence, a process that can be used to for examining laboratory work spaces for any LSD contamination. Assuming that there is no contamination, there will be no fluorescence from the LSD.

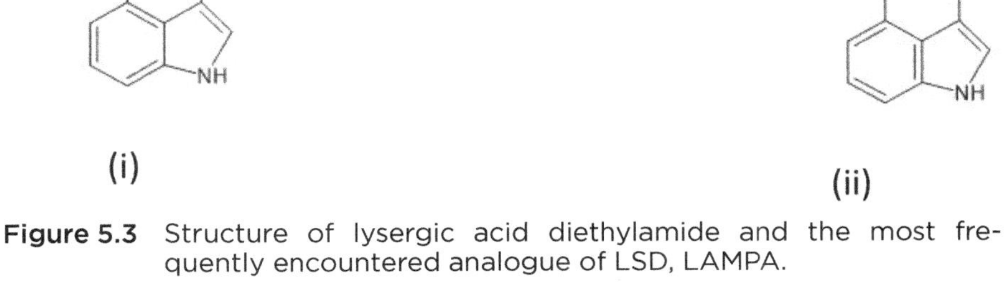

(i)   (ii)

Figure 5.3 Structure of lysergic acid diethylamide and the most frequently encountered analogue of LSD, LAMPA.

5.2.1 Physical Description of LSD

At the start of the analysis of LSD samples and blotter acids, a physical description should be made. This includes the number of dose units present, the number, size, shape and form of the dose units, and for blotter acids in particular, whether they are connected or separated, the nature, colour and distribution of any pattern or artwork on the dose units, whether or not the dose units are perforated and the nature of the perforations (whether any perforations pass right through the paper or partially so). Any packaging present should also be fully described. Such descriptions may allow the physical relationship with other LSD samples, packaging and cases to be established without the need, sometimes, for chemical analysis.

5.2.2 Sampling of Dose Units Expected to Contain LSD

The sampling of dose units expected to contain LSD is different to that where powders are seized. The United Nations Office on Drugs and Crime (UNODC) recommended method[1] is that where one sheet of paper is encountered, divided into a number of dose units, then that sheet is considered as one item and the assumption is made that the material is homogeneous. If a number of visually identical items are encountered and the number is up to 10, all the items should be sampled. For 11–27 items, 75% of the units should be sampled, which are chosen at random. Note that this differs from the standard UNODC protocol where N (the population size) lies between 11 and 100, the sample size (n) is 10. In cases expected to involve LSD, there are 28 or more items, the recommendation is to sample 50% of the units, chosen at random, with a minimum of 21 chosen and a maximum of 50.

At this stage and if the results from the screening tests are the same, the recommendation is that if the dose units are tablets, the selected items should be ground and sieved to form a powdered composite. Further, if the dose forms are gelatin, they should be combined and analysed as a whole. This does have the problem, however, that individual characteristics on individual dose forms will be lost, and so if the UNODC approach is adopted, the individual characteristics should be very carefully recorded prior to combining the materials together. If the results from the screening tests are not the same, the samples should be treated separately.

5.2.3 Presumptive Tests for LSD

Prior to the screening tests being undertaken, the drug should be extracted from the matrix in which it is contained into a suitable solvent (usually methanol) and the extract should be subjected to the screening test. For screening purposes, a small amount of methanol can be used where the target concentration to be achieved is approximately $1 \mu g \, ml^{-1}$. Whichever presumptive test is used, positive controls (LSD containing material) and negative controls (the extraction solvent) should be tested to demonstrate that the test is working, provide a reference colour reaction, and demonstrate that the reaction, which may be due to LSD, is due to the sample and no other source including the solvent in which the drug has been dissolved and extracted from its matrix.

There are two principal colour tests used for screening for LSD, namely Ehrlich reagent and van Urk reagent. Ehrlich reagent is a methanolic solution of *p*-dimethyl-aminobenzaldehyde (1 g in 10 ml) dissolved in orthophosphoric acid (10 ml). A small amount of the methanol extract of the drug is placed on a spot plate and two drops of reagent are added. The presence of LSD will result in a blue colour. However, the test is not specific and a number of indole alkaloids will result in a similar colour reaction. The reagent will also react with the tryptamine drugs psilocybin and psilocin (Section 5.3.3). A purple reaction indicates the presence of LSD and a pink reaction indicates dimethyltryptamine, respectively.

The van Urk reagent [0.125 g *p*-dimethylaminobenzaldehyde (DMAB) in 100 ml 65% sulfuric acid containing 0.1 ml 5% ferric chloride], also known as Hofmann reagent or *p*-DMAB-test solution, is used in a similar manner, again with both positive and negative controls. If LSD is present in the solution, a blue colour reaction occurs. However, the same colour is also obtained with the tryptamines psilocin and psilocybin, while dimethyltryptamine results in a yellow colour.

5.2.4 Thin Layer Chromatography of LSD and Related Drugs

Thin layer chromatography (TLC) of LSD can be carried out using relatively simple TLC systems. The stationary phase is usually silica gel containing a fluorescent indicator. Those systems that fluoresce at 254 nm are usually used. The sample to be applied to the TLC plate can be prepared with a target drug concentration of 1 mg ml^{-1}, with 2–3 µl (2–3 µg of drug) applied to the chromatographic plate. Recommended mobile phases are generally simple

– chloroform/methanol 9/1 by volume or chloroform/acetone 20/80 by volume. Positive (standard drug solutions) and negative controls (solvents alone) should always be applied to the TLC plate in addition to the samples. Once the chromatogram is developed, it should be observed under UV light prior to the use of visualisation reagents. Under UV light (254 nm), the compounds will appear as dark purple spots on a fluorescent background. Under long wavelength UV light (360 nm), these compounds typically fluoresce, giving a variety of colours depending upon the drug in question. Having visualised the chromatogram under UV light, it can be visualised with Ehrlich's reagent (Section 5.2.3). LSD and related compounds can be tentatively identified on the basis of comparison of their colour reactions and R_f value against known standards.

5.2.5 High-performance Liquid Chromatography of LSD

Most ergot alkaloids, including LSD, are thermally unstable, photolabile and do not occur in complex mixtures, and as a consequence, HPLC is a suitable method for their quantitative analysis and may be preferred over GC–MS.[2] Reversed-phase systems have been demonstrated to provide better separations. A typical system is recommended by the UNODC (Table 5.2). In this system, UV detection is at 313 nm. This does not provide selective detection because many benzene ring-containing systems absorb at this wavelength. These drugs are also an example of where fluorescence detection provides greater sensitivity and selectivity to the detection method. Selectivity is achieved because both the excitation and emission wavelength monitored in such detectors are specific to a particular drug. Sensitivity, reproducibility and repeatability of fluorescence detection are also enhanced by controlling the detection temperature of the flow cell – as temperature increases, sensitivity decreases. This is one of the reasons that most modern instruments have flow cell temperature controllers.

The problem encountered using HPLC to identify and quantify ergot alkaloids is that some co-elution occurs, for example, between LAMPA and LSD. Some authors have suggested that increasing the length of the HPLC column will address this issue, but this may cause the problem of band spreading as the column length increases. It is for this reason that methods with better resolution (GC–MS) or those capable of definitively identifying compounds in mixtures

Table 5.2 Exemplar HPLC conditions for the analysis of LSD using normal- and reversed-phase HPLC with UV and fluorescence detection.

Parameter	Setting
Normal phase	—
Column	Silica gel, 5 µm particle size, 125 mm × 4.6 mm i.d.
Mobile phase	10 mM ammonium perchlorate in MeOH, pH adjusted to 6.7 with 0.1 M NaOH
Flow rate	2.0 ml min^{-1}
Detection	UV at 313 nm
	Fluorescence: excitation 308 nm and detection 370–700 nm
Sample dissolution	Methanol
Injection volume	1 µl
Reversed phase	—
Column	Spherisorb 5-ODS or equivalent, 10 mm × 4.6 mm i.d.
Mobile phase	65% MeOH, 35% 25 mM disodium hydrogen phosphate in water, adjusted to pH 8.0 with 10% phosphoric acid
Flow rate	1.0 ml min^{-1}
Detection	UV at 280 nm
	Fluorescence: excitation 320 nm and emission 400 nm
Sample dissolution	Methanol
Injection volume	1–5 µl

(IR and NMR) are required, but thermal lability presents difficulties in terms of quantification of the drug.

5.2.6 Gas Chromatography–Mass Spectrometry of LSD

LSD and related drugs are particularly difficult to analyse by GC–MS. This is because these drugs are not thermally stable. They are also liable to react with residues in the injection port, column and detection system of the GC–MS and results in (i) poor chromatography, (ii) reduced sensitivity of the method of analysis and (iii) tailing of the compound during chromatography. Therefore, to successfully analyse this group of drugs, the GC–MS instrument must be scrupulously clean and free of contamination. Additionally, the samples should be protected from light prior to and during the analysis, including while being racked up on an autosampler, because the drugs are not fully photostable. The use of coloured (amber) sample vials is one way in which the decomposition of the drugs can be reduced. Alternatively, some laboratories wrap the vials in aluminium foil. This is also an example of an analysis where the use of deuterated analogues of LSD might be useful. The magnitude of the signal expected for a given amount of deuterated LSD (the internal standard for GC–MS analysis) would be known. If this

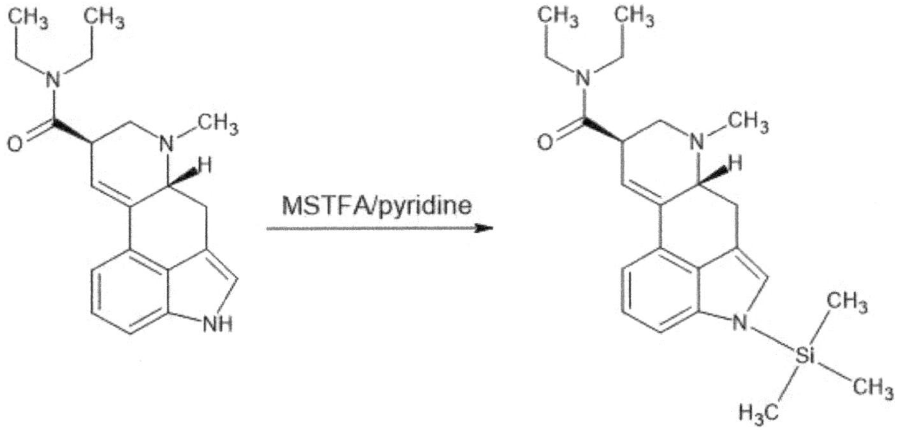

Figure 5.4 Reaction of LSD with MSTFA in pyridine to give the TMS derivative (TMS-LSD).

signal decreases in magnitude, then the suspicion must be that the deuterated internal standard is decomposing. As the deuterated LSD and LSD being identified and measured have the same chemistry, it must then be assumed that the LSD in the sample is also decomposing. Conversely, if the signal size remains the same, then the LSD in the sample may be deemed to be stable.

An alternative approach is to derivatise the drugs. It was demonstrated that N-trimethylsilyl-N-methyl trifluoroacetamide (MSTFA) could be used as the derivatisation reagent and that a MSTFA/pyridine mixture was a better derivatisation reagent (Figure 5.4) than MSTFA alone.[3] This is because the pyridine ensures that the drug is in its free base form and able to react with the derivatisation reagent. The reaction product, trimethylsilyl (TMS)-LSD, is a product from the reaction with the nitrogen in the indole ring system with the silylating reagent. As with many derivatives of drugs, TMS-LSD is more volatile, more thermally stable and less polar than underivatised LSD, resulting in better chromatography, greater instrumental sensitivity and improved mass spectroscopic information. Typically, ions are seen at $m/z = 395$ (71%) $[M]^+$, $m/z = 293$ (76%), $m/z = 253$ (100%) and $m/z = 279$ (41%).

5.2.7 Spectroscopic Methods for the Analysis of LSD

LSD is notoriously difficult to definitively identify and quantify. Spectroscopic methods provide advantages over chromatographic methods in that they can be definitive, spectral libraries can be developed for the identification and quantification of mixtures, and the samples can be recovered for other analyses and as reference compounds. IR and NMR spectroscopies have been successfully applied to LSD samples and although some of the techniques are quite old, they provide additional routes for the work-up of these difficult drug samples.

5.2.7.1 Infrared Spectroscopy of LSD

As it is difficult to definitively identify LSD by GC–MS, other techniques have been proposed for the identification of LSD and its isomers iso-LSD and LAMPA. One such method includes the use of preparative chromatography techniques followed by FTIR.[4] Identification can be achieved by the examination of the whole IR spectrum and especially the fingerprint region (750–1500 cm^{-1}). While such methodologies are relatively old, they do provide a mechanism for the clean-up and concentration of drugs from complex mixtures, dirty samples or those where the concentration is low. It is necessary, of course, to ensure that all of the equipment used, and solvents employed, are free of drugs. In the case of this group of drugs, this can be achieved by screening with UV light and looking for absorption and fluorescence. It is also important to note that the free base form of the drugs was examined in the study described. LSD sometimes occurs as the tartrate salt, increasing its solubility in water and improving the ability to handle the drug – much as in the case of amphetamines. The analyst will not know the salt form of the drug. It is therefore necessary to extract the drug from the paper sample using methanol, basify with ammonium hydroxide, dry and redissolve in methyl *tert*-butyl ether (MTBE) ahead of the preparative TLC used to clean up the sample of free base. The material is then worked up to provide the sample for IR. This can be achieved by separating a band placed along the origin of the TLC plate, rather than a spot, and developing the chromatogram. The band corresponding to that expected to be LSD can be identified under UV light and the stationary phase scraped off, typically with a clean spatula. The powdered silica gel can then be placed into a small tube, for example, a Pasteur pipette, plugged with cotton wool. The LSD can then be eluted from the silica gel using a polar solvent, for example, methanol. Once the sample has been prepared using preparative TLC and the subsequent work-up method, the drug is available for FTIR. LSD and related compounds can be identified on the basis of both the whole IR spectrum and the fingerprint region (750–1500 cm^{-1}) alone. It is, however, important to prepare reference spectra on the instrument being used because the quality of the spectra will depend upon (i) instrumental resolution, (ii) scan number (which affects signal to noise ratios) and (iii) the sample preparation including the matrix in which the sample is presented to the instrument.

5.2.7.2 NMR Analysis of LSD

A technique that offers the opportunity to overcome the thermal lability of LSD and also provides definitive identification of the drug is NMR spectroscopy, especially when one- and two-dimensional methods are used. NMR also offers the opportunity to determine the stereochemistry of LSD analogues. NMR has been applied to LSD and LSD analogues described as "research chemicals" in the recreational drug literature, including analogues that vary at the indole nitrogen (the N^1 position), including 1-acetyl and 1-propionyl-LSD and analogues differently substituted at the N^6 position [for example, N^6-allyl-6-norlysergic acid diethylamide (AL-LAD)] (Figure 5.5).[5,6] Samples extracted from their substrates can conveniently be dissolved in DMSO-d$_6$ with ^{1}H NMR spectra recorded at a field strength of 600 MHz and ^{13}C spectra recorded at 150 MHz. Application of both one- and two-dimensional techniques can lead to full structural determination of LSD-based drugs. DMSO is a convenient solvent for the NMR analysis of these drugs because it does not react with them, catalyse their breakdown or form artefacts. It is for this reason that NMR may be a useful technique for the analysis of LSD; because once the experimental data has been obtained, the drug can be recovered from the solvent in which it is dissolved.

5.3 Analysis of Fungal Material and Drugs Containing Psilocybin and Psilocin

Mushrooms of the genus *Psilocybe*, "magic mushrooms", are a source of the hallucinogen psilocybin. Interestingly, all but one of the hallucinogenic fungi (the ergots) are agaric fungi – those with an umbrella like cap on a stem (or stipe, as it is called).[1] The psilocybin content of fresh mushrooms of the genus *Psilocybe* varies between 0.2% and 3%, and an oral dose of 12–25 mg drug (600 mg–12.5 g fresh mushroom) is required to achieve a hallucinogenic response, with an onset time of 30 min and a duration of action of 3–6 h. The effects of these drugs

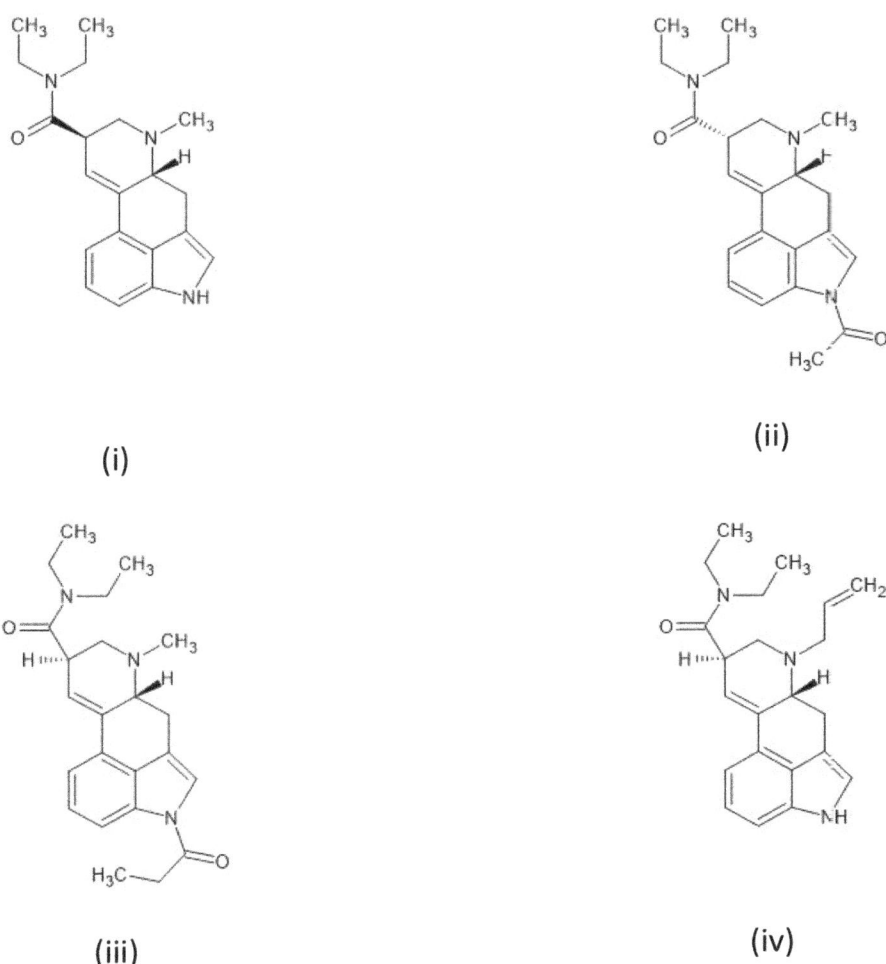

Figure 5.5 Structure of (i) LSD, (ii) 1-acetyl-LSD, (iii) 1-propionyl-LSD and (iv) 6-allyl-6-nor LSD.

are mainly due to psilocin, the major metabolite (dephosphorylation product) of psilocybin (Figure 5.1).[1] While a number of fungal genera produce psilocybin (*Panaeolus, Conocybe, Inocybe* and *Pluteus*), the most important species are *Psilocybe semilanceata* (Liberty Caps) and *Psilocybe cubensis*. Occasionally, edible preparations are encountered (Figure 5.6). These should be treated in the same way as fungal material.

 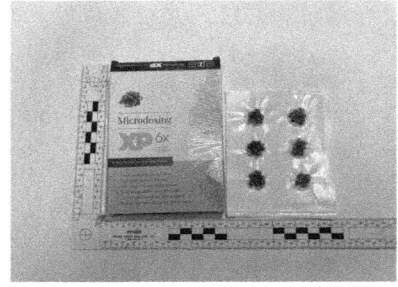

(i) (ii)

Figure 5.6 Examples of (i) whole dried fungal preparations and (ii) "edibles" manufactured from drug containing fungal material.

5.3.1 Packaging, Handling and Physical Examination of Fungal Samples

Seizing of magic mushrooms may include fresh, dried or "prepared" mushrooms, or powdered material, sometimes in capsules or tablets, which may be contained within one or more packages. Where the material is fresh, or moist, it is at risk of becoming mouldy on storage. To prevent this, the material should be weighed fresh, dried at 40 °C to constant mass and then re-weighed ahead of its storage as a dried specimen. This removes any water molecules from the materials and reduces the risk of mould developing. Once the materials are safely secured, they should be sampled according to the appropriate protocol of the jurisdiction in which the chemist is working. It is also known that psilocybin and psilocin are sensitive to exposure to air and light and so the materials should be appropriately stored – ideally desiccated in the dark.

5.3.2 Physical Examination of Drugs of Fungal Origin

Where an identification only is required, it may be possible for an expert mycologist to identify fungal material from whole or partial specimens, especially where the fruiting body is available. Mycologists can identify the material on the basis of the size and shape of the fungal material, along with its colour and texture. In addition, the shape of the gills, where the spores produced is important, as are the shape, dimensions and morphology of the spores themselves.[7] However, it should be strongly emphasised that such identifications for court-going purposes should only be made by experienced, expert, mycologists.

5.3.3 Extraction and Presumptive Testing of Mushroom-based Drugs

If quantitative analysis is to be undertaken, account should be taken of the variable water content of the fresh fungal material. To avoid difficulties associated with data interpretation, the material should be dried to constant weight at 110 °C, to thoroughly dry it out, and only then analysed. Prior to any chemical examination, where necessary the sample should be thoroughly homogenised using a pestle and mortar, a sieve, or commercial food blender. The problem with this procedure is that the whole sample/specimen is destroyed and all of the morphological characteristics are lost. It is also essential to ensure that whatever is used to homogenise the sample is free of contamination from the drug material before the homogenisation process is started. It may be preferable to take a small representative piece of the specimen/sample and homogenise that, leaving the remainder for subsequent examination and analysis.

Where the material is a powder or biological matter, a pre-extraction step may be necessary because coloured material may affect the interpretation of colour tests. The extraction can be carried out in methanol, bearing in mind the concentrations of drug likely to be found in the materials and the various sensitivities of the colour tests and instrumental methods to be used. Solid material can be removed using centrifugation or filtration, with all of the appropriate negative controls in place to demonstrate that any drug found came from the fungal material and nowhere else. The tests can then be applied to the extract obtained.

Two colour tests are commonly applied to fungal material, namely Ehrlich's reagent and Marquis reagent. The Ehrlich reagent (Section 5.2.3), used for indoles, will give a grey to grey/violet colour reaction with psilocybin and psilocin. The Marquis reagent (100 ml concentrated sulfuric acid mixed with 5 ml 40% formaldehyde) will give a yellow colour reaction. Both tests will give these colour reactions with a variety of drugs and drug classes. For example, Marquis reagent will also give a yellow/orange colour with amphetamine and methamphetamine. As a consequence, presumptive tests on mushroom material really should be regarded as a screening method only.

5.3.4 Thin Layer Chromatography of Drugs of Fungal Origin

TLC is a simple technique for the tentative identification of these drugs in the seized material. To prepare an extract for analysis from mushrooms, an extract at 10 mg mushroom/ml solvent can be prepared, filtered or centrifuged, blown down under a stream of nitrogen to approximately 0.5 ml and the extract spotted onto the TLC plate. Powdered mushrooms can be extracted at 1 mg powder per ml solvent with the extract being filtered/centrifuged and the extract analysed directly[1] (UNODC, 1989). Normal-phase chromatography can be used on silica gel plates containing a fluorescent indicator. Mobile phase systems include n-butanol/acetic acid/water 2/1/1 (by volume) and methanol/880 ammonia 100/1.5 (by volume). The plates should be dried before visualisation and care should be taken to ensure that all of the butanol has evaporated – it is a particularly difficult solvent to remove. The compounds can be visualised using UV light (254 and 365 nm), Ehrlich's reagent (prepared as for presumptive testing, Section 5.2.3), or p-dimethylaminocinnamaldehyde (0.5 g p-dimethylaminocinnamaldehyde in 50 ml methanol and 10 ml concentrated hydrochloric acid). The reagents give a range of grey violet/blue colour reactions with detection limits of 10–29 ng drug on plate. For all of the extraction processes and removal of solid materials, appropriate negative controls should be applied, including the full process minus the drug, to demonstrate that the apparatus and reagents used are free from any drugs. Positive controls should be used to provide reference colour and R_f data.

5.3.5 Gas Chromatography–Mass Spectrometry

Psilocybin is an example of a controlled drug that is not thermally stable and will thermally dephosphorylate under gas chromatographic conditions to psilocin. Derivatisation of the drugs is therefore required. Methods that have been used include derivatisation with bis(trimethylsilyl)trifluoroacetamide (BSTFA) and trimethylchlorosilane[1] and separation using GC–MS.[8] Separation can be carried out using simple gas chromatographic conditions with a 30 m × 0.25 mm i.d. HP-5 column (0.25 µm film thickness) with helium as the carrier gas (1 ml min^{-1}), splitless injection and 1 µl injections. An injection port temperature of 250 °C and column oven temperature programme of

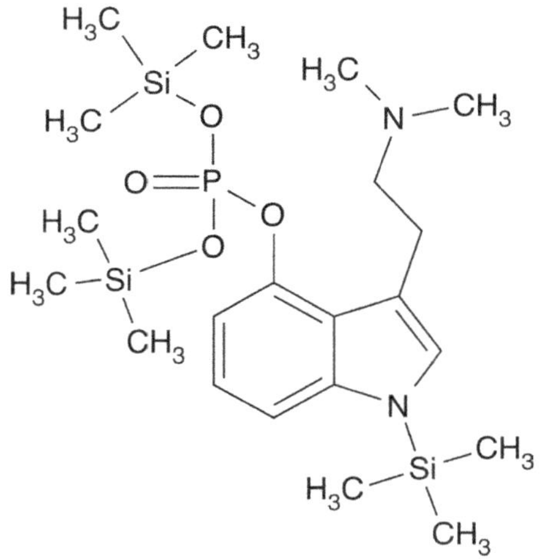

Figure 5.7 Structure of psilocybin 3TMS following derivatisation with BSTFA.

180 °C, rising at 20 °C min^{-1} to 320 °C with a 5 min hold can be used for effective separation.

The use of BSTFA, rather than other acid anhydrides produces a "cleaner" chromatogram for psilocybin 3TMS[8] (Figure 5.7). Ions at m/z = 485 [M − CH$_3$]$^+$, m/z = 455 [M − 3CH$_3$]$^+$ and m/z = 442 [M − Si(CH$_3$)$_2$]$^+$ are observed. Using this method, it is possible to both identify and quantify the psilocybin and psilocin in drugs derived from the parent fungal material.

5.3.6 High-performance Liquid Chromatography

HPLC is an analytical technique that is two orders of magnitude (*i.e.* 100 times) more sensitive for psilocybin and psilocin than GC–MS. Reported sensitivities include HPLC UV detection at 3.9 × 10^{-6} mol l^{-1} and HPLC with fluorescence detection 9.7 × 10^{-6} mol L^{-1}.[9] Isocratic normal phase, ion exchange, buffered reversed phase and ion pairing have all been applied, but are reported to require time-consuming extraction procedures, with low sensitivity and difficult matrix effects.[10] A one step extraction procedure using methanol on lyophilised (freeze dried) powdered material can be applied, and a reversed-phase HPLC system on a Spherisorb C-18 ODS-1 column, in a 0.3 M ammonium acetate pH 8.0/0.3 M ammonium acetate buffer delivered using a gradient programme and detection at 269 nm, also with diode array detection. The end capping of the HPLC column means that the drugs do not "stick" to the column that will result in tailing, improving the chromatography. Diode array detection led to absorption maxima being identified at 269 nm, 284 nm (sh), and 293 nm (sh).

Chemiluminescence detection has also been reported.[9] The analysis time was considerably shorter on a C_{12} column with a 95:5% by volume MeOH : 10 mM ammonium formate buffer at pH 3.5 at 5 min. Potassium permanganate/tris(2,2'-bipyridyl)ruthenium II was used for chemiluminescence detection along with UV detection at 269 nm. Sensitivity using the latter detection mode at 3.5×10^{-9} mol L^{-1} was achieved – three orders of magnitude better than using some other detection techniques, illustrating enhanced selectivity and sensitivity than that offered while using fluorescence and chemiluminescence detection methods.

5.3.7 DNA Analysis of Fungal Material

As with other drugs of botanical origin (for example, *Cannabis* products), when it is not possible to carry out morphological or chemical analysis of psilocybin-containing fungal material, it may be possible to carry out DNA analysis. Of particular use for such analyses is highly conserved DNA of low molecular weight, where there are tandem repeats of different sequences and numbers depending upon the fungal species being examined.[11,12] Of particular significance are the fungal genes for the 18S, 28S and 5.8S ribosomal RNA (rDNA), which occur in numbers of between 60 and 220 repeats in different fungal species. These genes are separated by internally transcribed spacer (ITS) and intergenic spacer (IGS) DNA. ITS RNA, from the transcribed ITS DNA during transcription and translation, is removed after the transcription of DNA. It is this DNA that contains polymorphisms that can be used for fungal identification. It is possible to extract DNA from dry fungal material and identify species. To extract the DNA, fungal material is powdered and then extracted in buffer (50 mM TRIS–HCl pH 8.0, 150 mM NaCl, 100 mM EDTA and 1% SDS). After the extraction period, 6M NaCl is added, followed by 10% CTAB/NaCl and then the mixture subjected to chloroform/isoamyl alcohol (24/1 by volume) extraction with overnight incubation. This precipitates the DNA. The DNA can then be redissolved and reprecipitated, purifying it yet further, and the RNA is removed followed by the reprecipitation of the DNA. The ITS-1 region is then amplified using ITS specific primers and the polymerase chain reaction. The resulting DNA can be sequenced and genus-specific sequences identified and used to prepare primers for testing of unknown samples. Amplification products can be separated using agarose gel electrophoresis. Using techniques such as this, it is possible to produce genus-specific ITS sequence data and from the primers specific to target genera. DNA from outgroup species (for example, the cultivated white mushroom *Agaricus bisporus*) does not result in amplification products when subjected to PCR using the specific primers. DNA analysis therefore provides a powerful analysis method in the drug chemists' arsenal for the identification of plants and fungi from which they are derived.

5.4 The Analysis of DMT

5.4.1 Introduction

N,N-Dimethyltryptamine (DMT) (Figure 5.1) is a recreational drug that is also found at endogenous levels in humans and a number of plants. There are a very large number of related drugs described in *Tryptamines I Have Known and Loved: The Chemistry Continues*.[13] DMT is used as a recreational drug and by various cultures and communities as a ritual drug. The dose, up to 50–60 mg, induces a psychedelic effect, and the drug acts as an entheogen, causing changes in perception, mood, consciousness, cognition and behaviour. The onset of the effects of DMT is rapid, inside 15 min, and the effects are short lived. It can be ingested, inhaled, or injected and the effects depend upon the mode of administration and the dose of the drug. They can be lengthened in the presence of monoamine oxidase inhibitors and are shorter lived if the drug is inhaled or injected when compared to oral ingestion. It is a compound found in Ayahuasca and is responsible for its hallucinogenic properties. Ayahuasca is a brew prepared from the *Banisteriopsis caapi* vine and *Psychotria viridis* leaves. While DMT is controlled, the brew itself is not. While a number of salt forms may be hygroscopic, users of the pure compound prefer the free base form because it has a lower boiling point. It is most frequently extracted from plant material rather than synthesised *de novo* because of the difficulty in obtaining starting materials and its difficult syntheses. In addition to the plants used to prepare Ayahuasca, DMT is also reported in *Mimosa* and *Acacia* species. At the street level, the drug may be encountered as crystalline material, powder, mixed with plant material or as part of a liquid drug.

5.4.2 Presumptive Tests for DMT

The Ehrlich (Section 5.2.3), van Urk (Hofmann) (Section 5.2.3), Marquis (Section 5.3.3) and Mecke reagents can all be applied to DMT. Mecke reagent can be prepared by dissolving 1 g selenious acid [H_2SeO_3, structurally $O=Se(OH)_2$] in 100 ml sulfuric acid. In the presence of DMT, the Ehrlich reagent will yield a pink/blue colour reaction, the van Urk a yellow colour reaction, the Marquis test an orange/brown reaction (as do amphetamine and methamphetamine), and the Mecke test a yellow/green reaction. Given the similar colour reactions to other drugs that contain indoles, and indeed other different drug classes, in order to identify and quantify DMT more sophisticated methods are required and colour tests should only be regarded as screening methods.

5.4.3 Thin Layer Chromatography of DMT

While there is a body of literature around the analysis of Ayahuasca, there is a dearth of literature on the analysis of street samples of DMT.[14] In one study TLC was carried out on silica gel that had been activated at 80 °C for 20 min which removes the water from the stationary phase and improves the chromatography. After application of the drugs to the chromatographic plates, the chromatograms were developed in methanol : ammonia 100 : 1.5 by volume, without saturation of the TLC tank. The ammonia ensures that the DMT is in the free base form during chromatography and reduces the amount of tailing of this very polar compound. The drugs were subsequently visualised under UV light (220 nm). The R_f value of DMT in this system was reported to be 0.5. The method was validated, determined to allow quantitative determination of the drug using a densitometer and UV absorption with an LOD of 3.3 μg and LOQ of 9.9 μg on chromatogram.

5.4.4 Gas Chromatography–Mass Spectrometry of DMT

As with thin layer chromatography, there is little literature on the use of GC–MS for the identification and quantification of street samples of DMT. However, there are methods described for the analysis of plant materials expected to contain DMT, and the methods used for the investigation of Ayahuasca preparations can be applied in an analogous way to the application of these methods to other drugs of herbal origin. DMT is also stable during the GC–MS analysis. Both GC–MS with quadropole detection and GC ion trap mass spectrometry (GC–IT-MS) have been applied to such samples. The use of GC is essential because of the need to separate the different components found in these complex mixtures with GC offering the required resolution to do so.

The application of straightforward GC–MS method has been described.[15] DMT can be identified on a GC–MS system using a split/splitless injector, a 5% phenyl-95% polydimethylsiloxane column (30 m × 0.25 mm × 0.25 μm film thickness), with high purity helium as the carrier gas with a flow rate of 1.2 ml min^{-1}. The oven temperature was 60 °C (3 min) rising at 8 °C min^{-1} to 200 °C, then rising to 280 °C (4 min) at 10 °C min^{-1}. The injection port temperature was 250 °C with splitless injections being made. The transfer line temperature was set at 280 °C, with a filament voltage of 70 keV, at 280 °C. Ions at m/z = 58 (iminium ion) and m/z = 188 (M$^+$) were monitored. The method allowed DMT determination in herbal samples (*Mimosa tenuiflora*) at a DMT concentration of 1.26–9.35 mg g^{-1}.

Similar methodology has been applied to beverages consumed during religious practices.[15] The system was used in splitless mode – whether or not this is necessary will depend upon the concentration of DMT in the sample. Separation was effected on a phenylmethylsiloxane column (30 m × 0.25 mm × 0.25 μm film thickness) with high purity helium as the carrier gas with a flow rate of 1 ml min^{-1}. The column oven temperature was 50 °C (3.5 min) rising to 280 °C at 20 °C min^{-1}, with a 10 min hold at the end of the analysis. Ion trap, ion source and transfer line temperatures were 220, 300 and 300 °C, respectively. The mass spectrometer was operated in a full scan mode in the mass range m/z = 40 to 400 at a scan rate of 0.6 scan per second. The dominant ion in the DMT mass spectrum is the iminium ion at m/z = 58. Other minor signal (intensity < 10%) fragments are observed at m/z = 188 [M]$^+$, m/z = 144, m/z = 130 and m/z = 115. Typically, an ion trap detector causes ions to be protonated so the iminium ion would be detected at m/z = 59. It is for this reason that the maximum sensitivity of the method is achieved by monitoring ions at m/z = 57 to m/z = 59. Identification and quantification can be achieved by comparing an authenticated drug with a commercial source.

5.4.5 High-performance Liquid Chromatography of DMT

HPLC can also been applied to the analysis of DMT, using a reversed-phase C-18 ODS-3 column with 5 μm particles, a pH 2.5 phosphate buffer in methanol (1:1 by volume), a 5 µL injection and UV detection at 280 nm. At this pH, an ion exchange separation mechanism effects the separation of the drug compounds. The material expected to contain the drug was extracted into methanol. It is also necessary to centrifuge or filter the extract to remove any particulate material ahead of the analysis to prevent it from blocking the analytical column. When determining whether or not a drug is present, reagent blanks and equipment blanks should also be analysed to demonstrate that they are free of drug contamination.

5.4.6 Nuclear Magnetic Resonance Spectroscopy of DMT

NMR analysis has been applied to DMT samples following extraction from plant material.[16] The ^{1}H NMR spectrum is relatively easy to interpret. A signal at $\delta = 2.35$ ppm is observed arising from the N-methyl groups. Two triplets are observed at $\delta = 2.70$ ppm and $\delta = 2.95$ ppm corresponding to the alpha and beta methylene groups, respectively. A multiplet is observed at $\delta = 7.00$ ppm, corresponding to protons 2, 5 and 6. Two doublets are observed at $\delta = 7.25$ ppm and $\delta = 7.50$ ppm due to protons 2 and 4 on the indole system. This is an example of a drug where the pH of the NMR experiment needs to be controlled – the acid/base work-up from the plant material containing the drug resulted in slight shifts in the NMR signals.

5.5 Analysis of Mescaline

Mescaline (3,4,5-trimethoxy-benzeneethanamine) (Figure 5.1) is a hallucinogenic compound derived from the peyote cactus (*Lophophora williamsii*). It occurs at concentrations up to 0.4% in fresh material and 3–6% in dried material. It is also known to occur in the San Pedro cactus (*Trichocereus macrogonus* var. *pachanoi*) albeit at a much lower concentration (0.8% mescaline in dried material). Of the hallucinogenic compounds described in this chapter, it is the only one that does not contain an indole group. Instead an ethylamine chain is attached to a benzene ring substituted with methoxy groups at carbons 3, 4 and 5. Despite that change, the psychoactive properties of mescaline are similar to other hallucinogenic drugs. The differences in chemistry will, however, impact on the detection and identification procedures used by a drug chemist, primarily because the presumptive tests, based upon the detection of an indole moiety, will not work for this drug. The peyote cactus can be consumed in different ways. These include ingestion as freshly peeled cactus, dried material and dried and ground powder.[17] It can also be brewed to prepare a tea/infusion of dark green colour.[17,18] The isolated drug usually appears as the sulfate or hydrochloride salt.[1] The greatest effect occurs around 2 h after ingestion and may last upwards of 8 h.

5.5.1 Packaging, Storage and Sampling of Cactus Material

As with all plant materials, it is essential to ensure that the cactus material is completely dried after seizing ahead of storage to prevent the material from becoming mouldy and the drugs decomposing. One way to achieve this is to air dry to constant weight at 40 °C. Dried plant material can then be subjected to chemical analysis after thorough homogenisation and powdering of the material. For quantitative analysis, some laboratories weigh the material fresh, as seized, and then dessicate the material to constant mass at 110 °C and then relate the amount of mescaline found to the original fresh weight of the plant material. In terms of sampling, the standard routines for choosing samples for analysis and preparing homogenised samples are recommended.

As with *Cannabis* plants, peyote (*Lophophora williamsii*) can be identified on the basis of botanical examination provided that enough whole plant material is available. However, unlike *Cannabis*, there are not enough microscopic characteristics (as is the case with the trichomes in the case of *Cannabis*) to make a definitive identification. It is therefore always best to carry out chemical analysis if botanical identification is in doubt.

5.5.2 Extraction of Mescaline

A number of extraction techniques have been used depending on whether the material being analysed is a plant material, a powdered drug, or an infusion prepared from the plant material.

5.5.2.1 Extraction of Mescaline from Plant Material and Powdered Drug

It is possible to extract mescaline directly from dried and powdered material using an extraction solvent of methanol–concentrated ammonia (99:1 by volume). The ammonia is present to ensure that the drug is in the free base form at the point of extraction. For TLC, the extract may be used directly, for HPLC- and GC-based analyses, at times an extraction step is performed with diethyl ether to remove lipids that may be present in the extract.[1] Being more polar and therefore relatively insoluble in the diethyl ether phase, the mescaline will remain in the methanol phase. Conveniently, methanol is compatible with both HPLC- and GC-based analyses where aqueous mobile phases are used, where diethyl ether is generally not.

5.5.2.2 Extraction of Mescaline from Aqueous Infusions Containing Mescaline

The analysis of mescaline in fusions used for medical purposes is also possible.[19] Biological samples (urine and hair) were also submitted, but the analyses of these toxicological samples are beyond the scope of this book. The analytical sample was prepared by means of liquid–liquid extraction (dichloromethane/isopropanol; 9:1, by volume) with a deuterated internal standard and the pH was adjusted to 9.2 with phosphate buffer and NaOH. Again, adjustment to this pH ensures that the drug is present in the less-polar free base form and is more soluble in the dichloromentane/isopropanol mix. The extract was then centrifuged for 10 min to remove solid materials, the organic phase collected and dried under a stream of nitrogen ahead of the instrumental analysis.

5.5.3 Presumptive Tests for Mescaline

Presumptive tests applied to these drugs should only be regarded as an indicator for the presence of mescaline in the material and definitive analyses should be considered obligatory. The Marquis test (Section 5.3.3) can be applied, where an orange colour develops. However, it should also be remembered that a wide range of drugs will produce similar colours with the Marquis reagent.

5.5.4 Thin Layer Chromatography of Mescaline

The extracts of material expected to contain mescaline can easily be analysed using thin layer chromatography on silica gel containing a fluorescent marker that fluoresces at 254 nm. Two simple systems can be used – chloroform/methanol/ammonia (82/17/1 by volume) or methanol/ammonia (100/1.5 by volume). In each case, the ammonia ensures that the mescaline is in the free base form and hence, the chromatographic problems associated with polar compounds and tailing are minimised.

Extracts prepared as described in Section 5.5.2 can be used for TLC analysis. Positive and negative controls should also be employed. Once the chromatogram has been developed, it should be air dried, visualised under UV light (254 nm), and developed with either fluorescamine (10 mg in dry acetone) or ninhydrin (10% in ethanol). Compounds that have same R_f values and colour reactions as the positive control can be tentatively identified as mescaline.

5.5.5 Gas Chromatography of Mescaline

It is possible to use the extracts prepared as above for simple GC analysis of mescaline. A simple system using a standard methylphenylsilicone column (DB-5 or equivalent), an injection temperature of 250 °C, a detector temperature of 310 °C and a column oven temperature of 150 °C (1.5 min) rising at 10 °C min^{-1} to 280 °C, with helium as the carrier gas and FID or MS detection can be used. However, since mescaline is a primary amine, unless the chromatographic system is scrupulously clean, it is likely that the mescaline will sorb to the surfaces in the instrument, either tailing very badly or not being detected. For this reason, and in analogous way to the amphetamines, derivatisation is recommended.

One reagent that has been used is pentafluoropropionic acid, which has been used to derivatise mescaline at 70 °C for 30 min.[19] Chromatography was carried out on an SPB-35 column (15 m × 0.25 mm × 25 µm), with an injection port temperature of 250 °C and splitless injection (1 min) and helium as the carrier gas at a flow rate of 1.0 ml min^{-1}. The column was programmed from 40 °C (1 min), rising at 40 °C min^{-1} to 200 °C (0.1 min) and then 15 °C min^{-1} to 290 °C (5 min). The mass spectrometer was set up with an ionisation source at 70 eV and ion trap detection in the scan range of m/z = 50–650. Ions were monitored at m/z = 181, 194, and 357 for mescaline pentafluoropropionic anhydride (PFPA). The samples could be both identified and quantified using this method.

References

1. Available from: https://www.unodc.org/pdf/publications/report_testinglsd_1989-01-01_1.pdf. Accessed 18 Aug 2024.
2. R. Gill and J. A. Key, *J. Chromatogr. A*, 1985, **346**, 423.
3. D. O. D. Costa, *Development and validation of a GC-MS method for identification and quantification of LSD in street seized samples*, Master's thesis, University of Coimbra, Portugal, 2020.
4. H. A. Harris and T. Kane, *J. Forensic Sci.*, 1991, **36**, 1186.
5. S. D. Brandt, P. V. Kavanagh, F. Westphal, A. Stratford, S. P. Elliott, K. Hoang, J. Wallach and A. L. Halberstadt, *Drug Test. Anal.*, 2016, **8**(9), 891.
6. S. D. Brandt, P. V. Kavanagh, F. Westphal, S. P. Elliott, J. Wallach, T. Colestock, T. E. Burrow, S. J. Chapman, A. Stratford, D. E. Nichols and A. L. Halberstadt, *Drug Test. Anal.*, 2017, **9**(1), 38.
7. R. Watling, *J. Forensic Sci. Soc.*, 1983, **23**(1), 53.
8. T. Keller, A. Schneider, P. Regenscheit, R. Dirnhofer, T. Rücker, J. Jaspers and W. Kisser, *Forensic Sci. Int.*, 1999, **99**(2), 93.
9. N. Anastos, S. W. Lewis, N. Barnett and D. N. Sims, *J. Forensic Sci.*, 2006, **51**(1), 45.
10. S. Borner and R. Brenneisen, *J. Chromatogr.*, 1987, **408**, 402.
11. J. C. I. Lee, M. Cole and A. Linacre, *Forensic Sci. Int.*, 2000, **112**(2–3), 123.
12. J. C. I. Lee, M. Cole and A. Linacre, *Electrophoresis*, 2000, **21**(8), 1484.
13. A. Shulgin and A. Shulgin, in *TIHKAL, The Continuation*, Transform Press, Berkeley, California, 1997.
14. B. Duffau, S. Rojas, S. Jofre, M. Kogan, I. Trivino and P. Fuentes, *J. Planar Chromatogr.--Mod. TLC*, 2014, **27**(6), 477.
15. A. A. Al Gaujac and S. Navickiene, *J. Chromatogr. B:Anal. Technol. Biomed. Life Sci.*, 15 January 2012, **881–882**, 107–110.
16. J. A. Fasanello and A. D. Plack, *Microgram J.*, 2007, **5**, 41.
17. C. Gambelunghe, R. Marsili, K. Aroni, M. Bacci and R. Rossi, GC - MS and GC - MS/MS in PCI mode determination of mescaline in peyote tea and in biological matrices, *J. Forensic Sci.*, 2013, **58**(1), 270–278.
18. J. H. Halpern, *Pharmacol. Ther.*, 2004, **102**(2), 131.
19. C. Gambelunghe, R. Marsili, K. Aroni, M. Bacci and R. Rossi, *J. Forensic Sci.*, 2013, **58**(1), 270.

6 The Analysis of Khat and Cathinones

6.1 Introduction

Khat (*Catha edulis*) is a shrub native to East Africa (Ethiopia, Djibouti and Kenya) and the Arabian Peninsula. It is a popular custom for people from these regions to chew fresh khat (*C. edulis*) leaves. The leaves are generally chewed fresh to achieve the maximum stimulant effect as cathinone, the main psychoactive compound, is reported to degrade with time after the leaves are picked and begin to dry out. It is for this reason that the leaves are wrapped and shipped in, for example, banana leaves, to keep them fresh. Stimulant effects similar to those caused by amphetamine are experienced by users. The effects of synthetic cathinones include euphoria, empathy, alertness, and talkativeness, with adverse effects such as paranoia, anxiety, panic attacks, hallucinations, excited delirium (extreme agitation and violent behaviour), depression, tremors and psychological dependence. The risk of overdose and self-harm associated with these drugs increases due to compulsive re-dosing owing to the psychological dependence that develops on these drugs. Another problem is that the purity of cathinones in street samples is not known, and as a consequence, there have been fatalities reported from unintentional cathinone overdose. These are due to the presence of cathinone and cathine, which are structurally similar to amphetamine (Figure 6.1).

Migrant communities introduced khat to Europe and North America where, in 2009, it became the second most popular plant-based drug, other than cannabis products, after *Salvia divinorum* (sage). Khat is

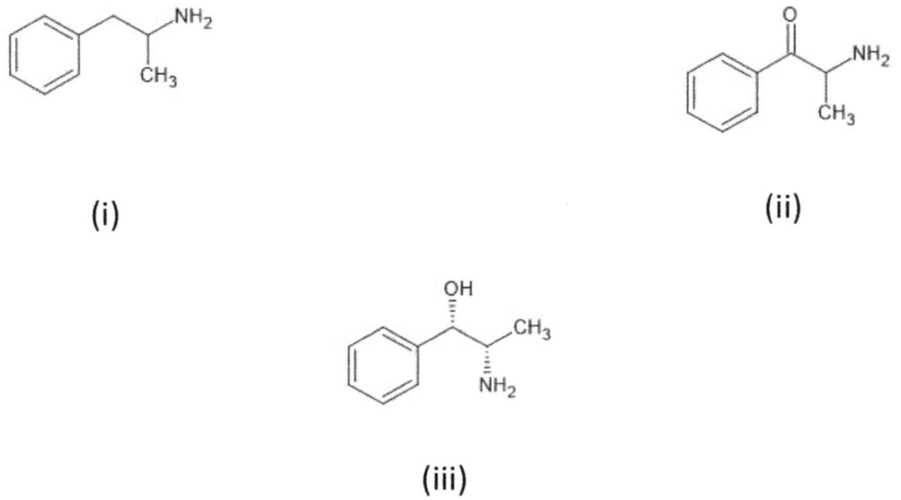

Figure 6.1 Structure of (i) amphetamine, (ii) cathinone and (iii) cathine.

known by different names worldwide, including chat, qut, qaad/jaad, miraa, mairungi, murungu, Arabian or Abyssinian tea and Muhulo.[1] Khat is available in different forms, ranging from a bundle of leaves and fresh shoots wrapped in banana leaves to dried infusions and an alcoholic extract. Based on the form of khat available, it can be chewed, ingested or smoked. A typical dose contains 100–200 mg of active compounds in one bundle of khat leaves.

Cathinones (exemplar compounds shown in Figure 6.2) are a class of designer drugs, based on cathinone, typically synthesised in clandestine laboratories, that are structurally similar to amphetamines. Cathinone is the principal active component present in the leaves of khat and is one of the drugs in this class that has a ketone group at the β-carbon atom in the amphetamine structure. For this reason, these drugs are considered β-keto analogues of amphetamine and are pharmacologically similar to amphetamines. Clandestine chemists began changing the parent structure of cathinone to enable the synthesis of new derivatives that are more potent than the parent drug and to circumvent drug legislation. Cathinone analogues are regularly modified by alkyl or halogen substitutions in the aromatic ring (*e.g.* mephedrone and clephedrone), methylenedioxy substitutions (*e.g.* methylone), benzyl substitution (*e.g.* 3,4-methylenedioxy-*N*-benzylcathinone) and pyrrolidinyl substitutions [*e.g.* 3,4-methylenedioxypyrovalerone (MDPV) and α-PVP]. Of the 138 cathinones currently monitored by the European Monitoring Centre for Drugs and Drug

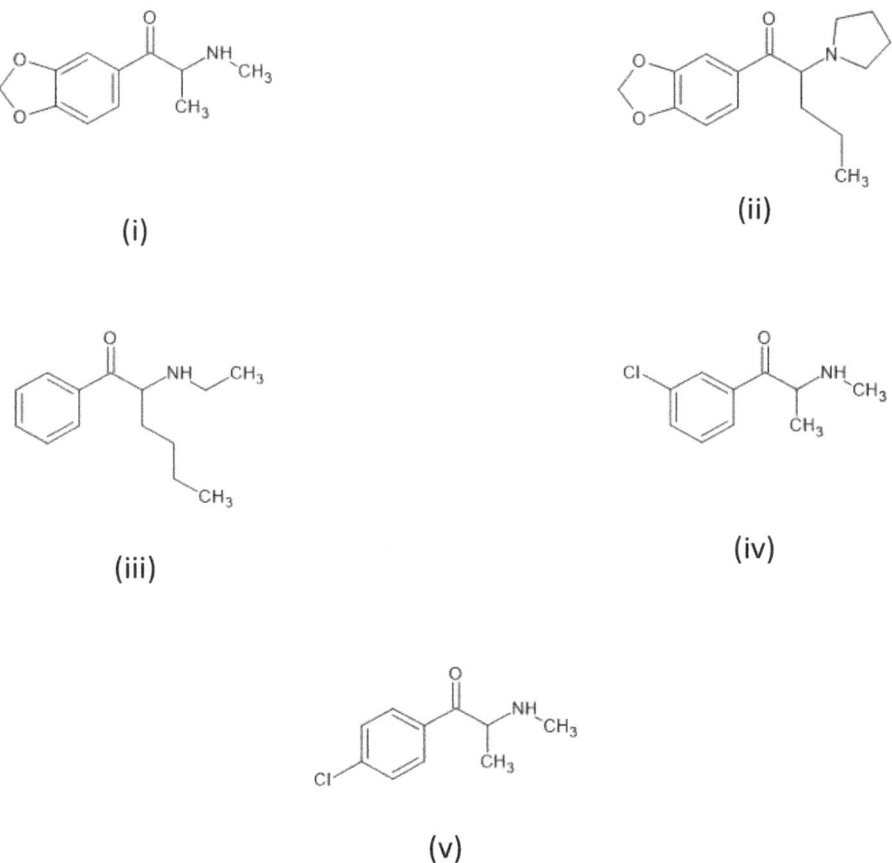

Figure 6.2 Examples of frequently encountered cathinones: (i) methylone, (ii) 3,4-methylenedioxypyrovalerone (MDPV), (iii) N-ethylhexedrone, (iv) 3-chloromethcathinone (3-CMC) and (v) 4-chloromethcathinone (4-CMC).

Addiction (EMCDDA), mephedrone, methylone, MDPV, N-ethylhexedrone, 3-chloromethcathinone (3-CMC) and 4-CMC (clephedrone) have gained international popularity as drugs of abuse. Synthetic cathinones are sold under a wide variety of brand names; examples include Bliss, Cloud Nine, Lunar Wave, Vanilla Sky and White Lightning.

The drug group can be considered to be amphetamine-type analogues. For example, cathinone is similar to amphetamine, methcathinone to methamphetamine, mephedrone to 4-methylmethamphetamine, and clephedrone to chloromethamphetamine (Figure 6.3). Cathinones have both ketone and amine functional groups and therefore they may undergo keto–enol tautomerism. As shown in Figure 6.4, tautomers are isomers of a compound that differ in the position of the protons and electrons, and tautomerisation involves intramolecular proton transfer, in this case, in mephedrone. This is something that forensic chemists need to be aware of as it can lead to,

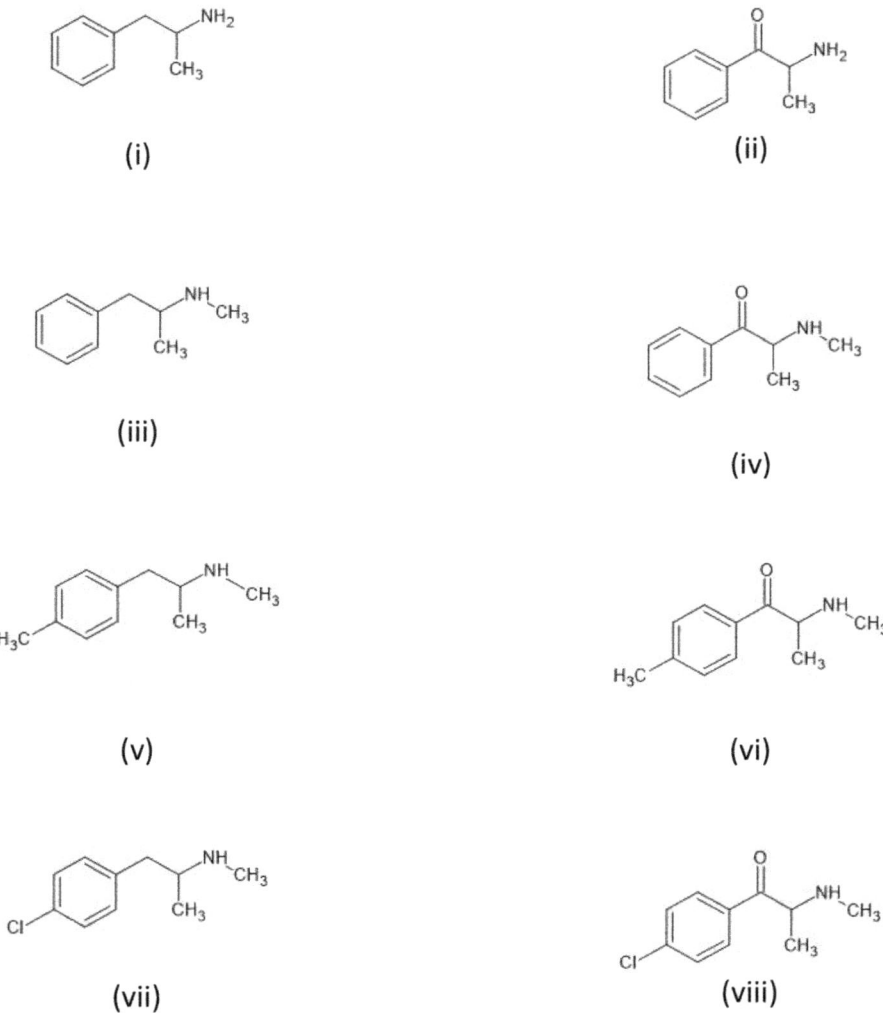

Figure 6.3 Examples of structural similarities between amphetamines and cathinones including (i) amphetamine and (ii) cathinone, (iii) methamphetamine and (iv) methcathinone, (v) 4-methylmethamphetamine and (vi) mephedrone and (vii) 4-chloromethamphetamine and (viii) 4-CMC (clephedrone).

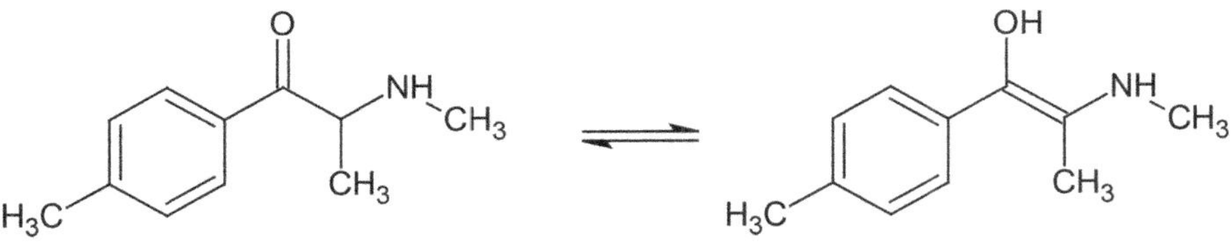

Figure 6.4 Tautomerism of mephedrone, an example of tautomerism encountered in cathinones.

for example, underestimation in quantification and could, potentially, lead to the misidentification of the drug.

In the cathinones, the carbon atom adjacent to the amine is stereogenic (*i.e.* chiral), giving rise to two enantiomers – the *R* and *S* isomers using *R/S* notation. Therefore, pharmacologically, two distinct forms (*i.e.* enantiomers) of each cathinone are possible. In illicit preparations, racemic mixtures (mixtures of both the *R* and the *S* isomer) are likely to be present. Clandestine manufacturers have identified this loophole in legislative systems and frequently synthesise isomers that are not controlled. This can be achieved by choosing starting materials for synthesis with the correct stereochemistry. However, not all isomers are equally potent. For example, *S*(−) methcathinone is more potent than the racemic mixture of methcathinone, which is more potent than *R*(+) methcathinone[2] (Figure 6.5).

Legislative control of the cathinones is complex and depends on the country where the analysis is being carried out. For example, in the UK, cathinones are generally Class B or Class C drugs, while in Australia, for example, some cathinones are controlled as prescription drugs, while others are prohibited (Table 6.1). There is, however, some correlation between the scheduling of the drug internationally under the UN convention and control at the national

Figure 6.5 Structure of (i) *R*(+) methcathinone and (ii) *S*(−) methcathinone.

Table 6.1 Illustrative examples of control of cathinones under different legislative systems (UK – Misuse of Drugs Act 1971, USA – Controlled Substances Act 1971, Australia – Therapeutic Goods Act 1989 and United Nations – Convention on Psychotropic Substances 1971).

Drug	UK	USA	Australia[a]	United Nations
Cathinone	Class C	Schedule I	Schedule 9	Schedule I
Cathine	Class C	Schedule IV	Schedule 4	Schedule III
Mephedrone	Class B	Schedule I	Schedule 8	Schedule II
Methylone	Class B	Schedule I	Unscheduled	Schedule II
MDPV	Class B	Schedule I	Schedule 9	Schedule II
N-Ethylhexedrone	Class B	Schedule I	—	Schedule II

[a]Australia (Schedule 4 – prescription only, Schedule 8 – controlled drug, Schedule 9 – prohibited substance)

level through which the UN legislation is implemented. This is one of the reasons that careful analysis of drugs thought to contain cathinones must be undertaken to correctly identify the drug.

6.2 Forms, Dosage and Effect of Synthetic Cathinones

Synthetic cathinones are frequently encountered as white, off-white, yellowish or brown powders in both amorphous and crystalline forms. Liquids and tablets are also available in the illicit drug market.

Cathinones are taken in a manner similar to amphetamines. For example, mephedrone can be insufflated (snorted), injected, orally ingested or administered rectally. Ring-substituted cathinones (*e.g.* methylone and butylone) are sometimes snorted in a similar fashion to cocaine. A typical illicit dose varies from 3 to 250 mg and is based on the potency of the individual cathinones, mode of administration and tolerance of the user (Table 6.2). MDPV (a cathinone with a pyrrolidine ring) is reported to be one of the more potent cathinones (a dose of 3–20 mg is usual). The increased potency is due to the more lipophilic nature of the molecule, and hence cell membrane permeability, owing to the presence of the pyrrolidine ring and tertiary amine. The more lipid soluble the drug, the more rapid the onset time.

Table 6.2 Exemplar data for single doses of cathinones, mode of administration, onset time and duration of effect.

Drug	Administration route	Typical dose (mg)	Onset time (min)	Duration of effect (h)
Methcathinone	Oral	60–250	>5	1–1.5
	Intramuscular			
	Intravenous			
Mephedrone	Oral	100–250	45–120	2–4
	Intramuscular	5–75		0.5–1
	Intravenous	50–75		0.25–0.5
Methylone	Oral	100–250	30–45	2–5
MDPV	Oral	3–20	15–30	2.5–3
	Intramuscular	—	<30	6–8

6.3 Identification of Khat and Cathinones

6.3.1 Identification of Plant Material as Khat (*Catha edulis*)

Botanical and microscopic analysis of khat is possible by a competent botanist. It is a slow-growing plant that can reach up to 25 m in height. The leaves are oval and finely toothed/serrated. The flowers, with five whitish petals, are formed on stalks in the axils of the leaves. The young stems may be red in colour. However, post-harvest, some of these botanical features may be lost, and it is better to carry out a chemical analysis to be certain of the identity of the material.

The detection of cathinone and cathine within the plant material is required in the laboratory for the definitive identification of plant material. To extract these drugs, 5 g of seized material can be crushed using 100 ml of 0.1 M HCl followed by filtration to remove the solid plant material. The acid solution is then basified using 1 M NaOH and extracted 3 times each with 50 ml of chloroform. The organic layer, which extracts the drug from the mixture in the free base form, is combined, filtered and evaporated to dryness under nitrogen. The residue is then dissolved in 1 ml of chloroform and subjected to GC–MS analysis using the same instrumental parameters as for synthetic cathinone (Section 6.3.2.5). Cathinone is unstable in plant material and in solution in the extract. It can be reduced to (+)norpseudoephedrine and (−)norephedrine or oxidised

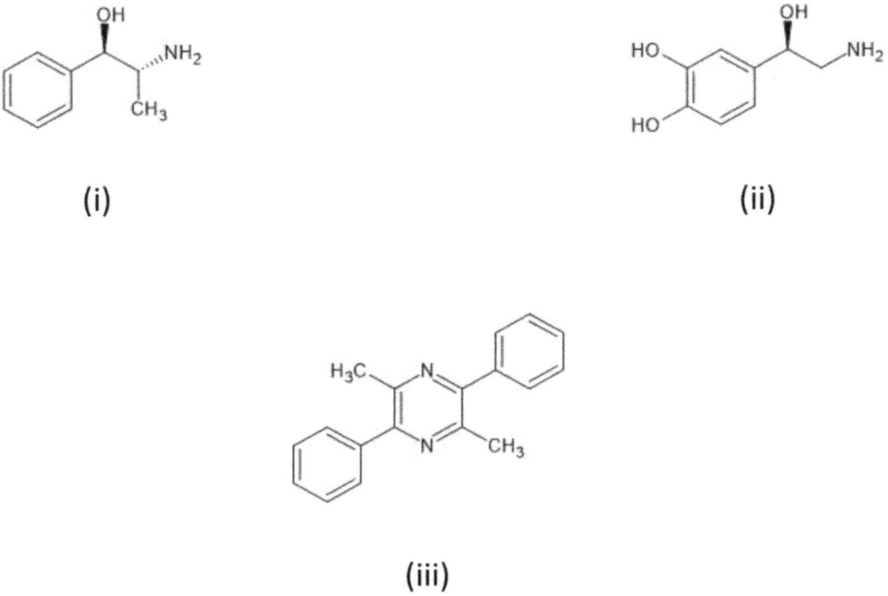

Figure 6.6 Structure of (i) (+)norpseudoephedrine, (ii) (−)norephedrine and (iii) 2,5-dimethyl-3,6-diphenylpyrazine.

to a dimer (2,5-dimethyl-3,6-diphenylpyrazine) (Figure 6.6). All of these can be detected in the extract. This means that the extract, if the drugs are not chemically derivatised, which stabilises the parent compounds, should be analysed as soon as possible after preparation. The decomposition of the cathinones also makes a comparison of the plant material thought to be khat problematic because it cannot be determined whether differences in chemical composition are due to genuine differences in the plant material or due to different rates of chemical decomposition of the drugs.

6.3.2 Identification of Synthetic Cathinones

6.3.2.1 Sampling and Physical Description

Cathinones might be encountered as powder samples or tablets similar to amphetamine or ring-substituted amphetamine (ecstasy)-type drugs. Therefore, a forensic chemist should develop a sampling strategy on the basis of the request received, the analysis required and the jurisdiction in which they are operating (Chapter 2). Due to the cost and time associated with analysis, it is a good practice to determine the sample population and have a sampling plan in place to reduce the number of samples required for the analysis without compromising the confidence level in the results. Before extraction, the samples should then be homogenised and a small amount of the material is used for further extraction and analysis.

6.3.2.2 Colour Tests

Of the presumptive (colour) tests available, the Zimmermann reagent has been widely applied to test for the presence of cathinones in drug samples. The Zimmermann reagent has two component solutions. The first is a solution of 1% w/v 1,3-dinitrobenzene in methanol and the second is 15% w/v potassium hydroxide in water. One drop of each, in sequence, is added to the test substrate, and the colour reaction is observed. A wide variety of colour reactions occur when cathinones react with the Zimmermann reagent (Table 6.3). However, there are also a number of other drug classes that react, and so it is necessary to perform a range of colour tests, as described throughout this text, on any drug seizure to fully understand the drug content.

6.3.2.3 Microcrystal Tests for Cathinones

Microcrystal tests indicate the presence of a particular drug class by the formation of crystals with specific morphologies and optical properties when a reagent or several reagents are added dropwise to the substance being tested. They are quick and simple, as are presumptive tests, but far more sensitive. Microcrystal tests have been developed to detect selected new psychoactive substances (NPS) including mephedrone.[3] Equal volumes (10 µl) of aqueous mephedrone and the test solution (mercury chloride, 10 g l^{-1} in water) are allowed to react on a glass slide, which results in crystal formation (paddlewheels or rosettes of blades). The crystals are then viewed under a polarising microscope. A photograph of crystals

Table 6.3 Exemplar colour reactions of cathinones with the Zimmermann reagent.

Drug	Colour reaction
Butylone	Pink
MDPV	Yellow
Mephedrone	Purple
Methcathinone	Purple
Methylone	Purple
Naphyrone	Yellow

formed with standards (tested/treated in a similar manner) should then be compared with the sample. At the time of writing, microcrystal tests have not been reported for other cathinones.

6.3.2.4 Thin Layer Chromatography of Cathinones

TLC is a convenient method for screening and comparing drug samples, including cathinones. The UNODC TLC method[4] recommends silica gel containing a fluorescent indicator as the stationary phase, with a mobile phase of ethyl acetate/methanol/25% ammonia (85/10/5 by volume). Ammonia is added to ensure that the drugs, which are themselves highly polar, are in the free base form, improving their chromatographic properties and behaviour. Standards should be dissolved in methanol at around 1 mg ml^{-1} and street samples dissolved to give an approximately similar concentration. Between 1 μl and 5 μl of these solutions should be applied to the chromatographic plate, the applied spots dried and the chromatogram developed. Once developed, the chromatogram should be dried and visualised under UV light (254 nm) and with a ninhydrin reagent (typically 2% w/v in acetone). Under UV light, any compounds that absorb UV light, including cathinones, will appear as purple spots against a bright (usually yellow) background. If using ninhydrin, the chromatogram should be heated at 80 °C for up to 40 min to fully visualise the separated compounds that react with ninhydrin, forming blue, black and purple products. One of the problems associated with the TLC analysis of cathinones is that the data obtained demonstrate that many of the R_f values and colour reactions are very similar. It is for this reason that the TLC of cathinones should only be used as a screening method at best and more sophisticated methods of identification are required for confirmatory analysis.

6.3.2.5 GC-MS Analysis of Cathinones

Synthetic cathinones can be identified using GC–MS. The first steps are sample preparation, which includes (i) dissolution (including the selection of the correct solvent and removal of any insoluble material by filtration or centrifugation), (ii) selection and addition of an internal standard and (iii) derivatisation reactions.

Cathinone samples are encountered as tablets, capsules, powders and even liquids. Most of the active compounds present in seized samples are soluble in organic solvents, which makes sample preparation easy. Samples can be prepared at 1 mg ml^{-1} in methanol.

An appropriate amount of internal standard can also be added at this stage. If calibration data are to be used to establish the linear range of the response of the instrument, a wide range of concentration points should be included that bracket and cover the anticipated concentration of the drug in the sample, with the sample ideally in the middle of this concentration range. For street drug analysis, concentration points should include both the high (mg ml^{-1}) and low (µg ml^{-1}) concentrations, as cathinones encountered in street samples can have widely differing purity levels. At least triplicate analysis is recommended as a good practice to enable statistical analysis.

The internal standard is a known compound selected by the analyst and is added at the same concentration and volume to all standards and samples. Internal standards should be stable and not react with the ketone and amine moieties present in cathinones. It should not already be present in the sample (*i.e.* other possible drugs and adulterants mixed with the cathinones). A compound that shares similar physical and chemical properties to the analyte is considered suitable. On this basis, deuterated analogues of cathinones are considered best (*e.g.* mephedrone-D_3, naphyrone-D_5, MDPV-D_8). However, deuterated analogues can be expensive and are not always available commercially. It is for this reason that *n*-alkanes are commonly used as internal standards. However, deuterated analogues have the advantage of acting as check standards. These standards give a known signal size in the detector. Should the signal size decrease, the deuterated compound decomposes, and it can therefore be assumed that the drug of interest also undergoes the same decomposition reactions. Thus, check standards are a good means of ensuring that samples are not decomposing while being queued for analysis on an instrument.

There are various peer-reviewed chromatography and mass spectrometry-based methods available for the analysis of cathinones. Whichever method is used, it is important to ensure that the method is validated and allows for the separation of possible cathinones and adulterants present in the street sample. A method recommended by the UNODC[4] allows good separation of cathinones and adulterants. The oven conditions are 90 °C for 1 min, increased to 300 °C at 8 °C min^{-1}, and held isothermally at 300 °C for 10 min. A phenylmethyl silicone column (BP-5 or equivalent) can be used conveniently (column dimensions 30 m × 0.25 mm i.d. × 0.25 µm film thickness). The injection temperature is set to 225 °C, with He as the carrier gas flowing at 1 ml min^{-1}. Full scan electron impact mass spectrometry is used to detect the cathinones in total ion count mode between 50 and 500 atomic mass units. The GC interface is set at 300 °C, ensuring that there is no condensation of the analytes in the interface, even at the highest column temperature; the source temperature is 230 °C and the quadrupole temperature is 150 °C. This method starts the temperature programme at a lower temperature than most methods, with the oven programme starting at 90 °C, and incorporates a slower temperature ramp rate (8 °C min^{-1}), which can offer improved analyte resolution but takes longer to complete than many published methods. Additionally, many methods, including the method described, recommend a high-temperature hold at the end of the temperature programme in order to "clean" the column. Continuous exposure of the column to high temperatures (300 °C) with long hold times may decrease the life of the column in addition to resulting in column bleed, which interferes with the chromatogram and reduces sensitivity. During street drug analysis where drug concentrations could be higher, the use of sample splits is recommended to avoid detector saturation. Split ratios in the range of 1 : 10–1 : 20 can conveniently be used.

Identification of unknown cathinones present in street samples *via* GC–MS is achieved by comparing retention time (or retention index against an internal standard) and the mass spectrum of the analyte with that of the standard analysed using the same method. The NIST library, or any other commercially available library, can be used at an initial stage for tentative identification. However, this cannot be used for confirmation purposes. Using the mass spectrum of MDPV (Figure 6.7), it is straightforward to understand the fragmentation pattern of the molecule. MDPV (RMM 275) is fragmented/ionised during the mass spectrometry ionisation unit to form ions at m/z = 274 $[M-H]^+$, m/z = 149 and m/z = 126 (100%).

Thermal degradation of cathinones has been reported when analysed using GC-based techniques. To avoid the formation of the artefacts, the use of a deactivated liner and a clean column is recommended. Once unknown compounds are identified and confirmed, the next stage is quantification. Traditional GC–MS techniques are used for the quantification of cathinones.

6.3.2.6 HPLC and LC–MS Analysis of Cathinones

Liquid chromatography is another separation technique that can be used for forensic analysis of cathinones. The most commonly used HPLC system for cathinones consists of a non-polar reversed-phase

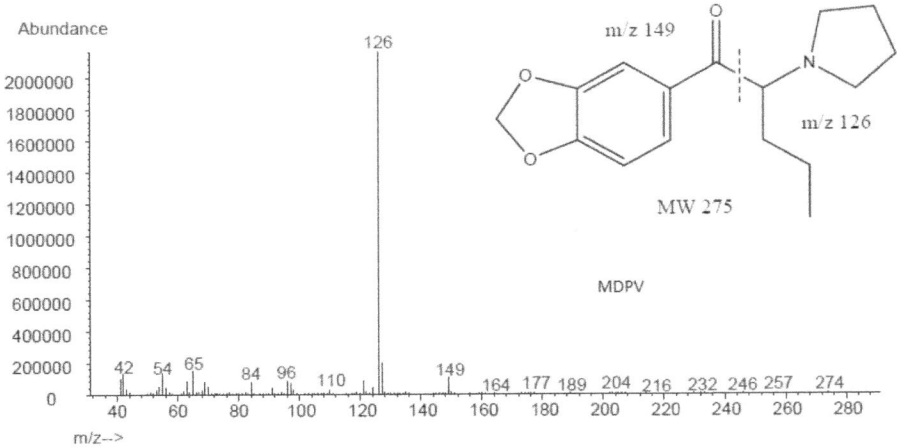

Figure 6.7 Mass spectrum and fragmentation of MDPV under conditions of electron impact mass spectrometry.

column. In comparison to GC analysis, HPLC does not require the addition of an internal standard because the injection volume on an HPLC instrument is fixed. HPLC is also a better technique for the analysis of polar compounds. Different detection systems are coupled to HPLC, but the most commonly used are diode array detection (DAD) and MS detection. As solvents and water grades are cheaper for DAD than those needed for LC–MS (*i.e.* LC–MS grade), HPLC–DAD is generally used during the method development stage. A typical system effects separation on a reversed-phase column although the particle sizes are smaller than those of conventional columns (3 µm against the normal 5 µm) (C-18, 150 × 4.6 mm i.d., 3 µm particle size). Smaller particle sizes can improve the resolution and chromatography of analytes, but the back pressure must be monitored to ensure that it does not rise too high and crush the top of the column, which in turn results in poor chromatography and peak shapes. The sample size is 10 µl, and an internal standard of nicotinamide can be used. The mobile phase can be 28 : 72 (by volume) methanol : 10 mM ammonium formate adjusted to pH 3.5 with formic acid, delivered at a flow rate of 0.8 ml min^{-1}. Detection is with a diode array detector.

When MS is coupled with LC, it is then a confirmatory technique used as an alternative technique to GC–MS for the analysis of polar and thermally labile compounds. Drug stock solutions can be prepared in methanol (1 mg ml^{-1}) and diluted in the HPLC mobile phase. An internal standard is not required for HPLC–DAD analysis, but researchers have started adding internal standards during LC–MS analyses. A good internal standard is one that is structurally similar to but distinct from any drug likely to be found in casework samples. Additionally, it will neither react with the drug being analysed nor catalyse the drug to break down. Samples should be prepared in a manner similar to GC analysis, except for the replacement of the solvent with the mobile phase to be used for LC sample preparation.

While developing LC–MS-based methods, analysts need to consider sample dissolution, column selection, mobile phase composition and pH adjustment. To overcome possible ionic interaction between the basic cathinones and silanol groups on the column, the mobile phase pH should be adjusted to around pH 2 by adding 0.5% by volume formic acid in the mobile phase. In a similar way to GC–MS analysis, the identification of unknown cathinones present in street samples *via* LC–MS is achieved by comparing the retention time and the mass spectrum of the analyte with that of the standard analysed using the same method. Commercial databases/libraries are not available for LC–MS analyses as results vary between labs. It is for this reason that individual labs need to create their in-house databases. When using LC–MS techniques, protonated molecular ions $[M+H]^+$ as base ions are detected. Additionally, LC–MS usually results in less fragmentation than GC–MS, which may result in difficulty for confirmation of unknown compounds. Because there is less fragmentation, there may be too few a number of different ions to reliably identify a compound. The use of at least 3 ions (the base ion/molecular ion and 2 other confirmation ions) is recommended for confirmation. Some authors have used two different columns for confirmation purposes, combining two sets of retention time data with the mass spectroscopic information.

6.4 Other Analytical Techniques

Cathinone samples have been analysed using multiple techniques. This is partly due to the unavailability of reference standards at the time a drug appears in the market. This means that the analyst may need to synthesise standards and characterise them. FTIR (Fourier transform infrared spectroscopy), ^{1}H, ^{13}C and ^{19}F NMR (nuclear magnetic resonance spectroscopy), X-ray crystallography and Raman spectroscopy have been used for this purpose.

FTIR provides a unique spectrum for each compound. A comparison and confirmation of the identity is possible. However, for street drugs when purity is unknown, identification and confirmation might not be possible unless GC coupled with IR is used in order to separate the cathinone from the other components of the drug matrix. Alternatively, some laboratories develop libraries of mixtures of compounds that are found to contain cathinones, and these mixtures are used to provide reference spectra to identify the components of drug mixtures at street level that have been seized. FTIR spectra of powder samples can be obtained by using the KBr disc method or an ATR-FTIR. FTIR is popular because it requires little or no sample preparation, provides a non-destructive analysis, is quick and easy to use and is portable. However, for confirmation, the sample spectrum needs to be compared with the standard spectrum obtained using the same method. When used in combination with other methods, FTIR adds value to the investigation process.

However, to elucidate molecular structure and to establish purity, NMR and X-ray crystallography-based techniques are used. These are generally used when standards are synthesised and characterised rather than in a routine casework. X-ray crystallography is useful for identifying the emerging unknown designer drugs and their enantiomeric forms. This method is rapid and useful if crystal particles are present in the samples. Both of these techniques have been used in the discrimination of positional isomers (*e.g.* 4-MMC, 2-MMC and 3-MMC) and enantiomers, which is often the case with cathinones.

6.5 Profiling of Cathinones

In order to understand cathinone profiling and the impurities likely to be encountered and to establish a link to the batch or the method of synthesis or the place of origin, it is important to have a clear understanding of clandestine synthesis. Clandestine laboratories producing synthetic cathinones have been reported in Estonia, Italy, the Netherlands and Poland. Due to cheap, easily available and replaceable precursors used to synthesise cathinones, the clandestine drug market continues to evolve with minor modifications of drugs designed to circumvent legislative systems.

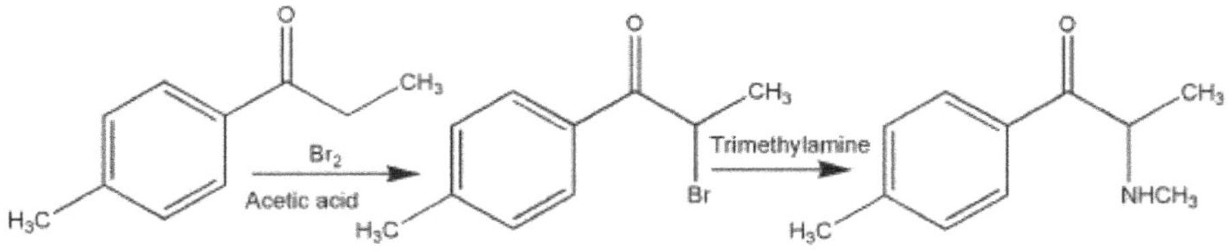

Figure 6.8 Synthesis of mephedrone from 4-methylpropiophenone.

The cathinone analogues are high-purity products, and there is little published information available about their impurity profiles. Synthesis methods include (i) selection of the appropriate β-ketoarylalkane and (ii) bromination, followed by (iii) an amination reaction using the desired amine. As an example, the synthesis of mephedrone involves the acid-catalysed bromination of 4-methylpropiophenone followed by the reaction with intermediate trimethylamine to form 4-methylmethcathinone or mephedrone (Figure 6.8).

These reaction intermediates and associated reaction by-products might be expected in samples where chemical impurity profiling has been undertaken. The use of inorganic constituents or stable isotopes, for example, is an area for further research.

References

1. L. Gautam, A. Shanmuganathan and M. D. Cole, Forensic analysis of cathinones, *Forensic Sci. Rev.*, 2013, **25**(1–2), 47.
2. R. Young and R. A. Glennon, Discriminative stimulus effects of S(−)-methcathinone (CAT): a potent stimulant drug of abuse, *Psychopharmacology*, 1998, **140**, 250.
3. L. Elie, M. Baron, R. Croxton and M. Elie, Microcrystalline identification of selected designer drugs, *Forensic Sci. Int.*, 2012, **214**(1–3), 182.
4. Available from: www.unodc.org/documents/scientific/STNAR49_Synthetic_Cathinones_E.pdf. Accessed 18 Aug 2024.

7 The Analysis of Opiate Drugs and Heroin

7.1 Introduction

Heroin is a drug that is widely abused around the world with estimates, at the time of writing, of more than 15 million people across the globe consuming drugs (opium, morphine and heroin) derived from the opium poppy, *Papaver somniferum*. In the United Kingdom, heroin refers to a mixture of products (Figure 7.1) formed from the synthesis of heroin from opium and morphine or carried over in heroin production, while in the United States, heroin is used to mean the single compound, diamorphine. It is for this reason that the forensic scientist should be very clear about the use of their language around this drug. In this chapter, the term heroin will be used to mean the mixture of drugs that occur when heroin is synthesised from morphine and then mixed with adulterants (pharmacologically active, for example, caffeine and paracetamol/acetaminophen) and diluents (pharmacologically inactive, for example, starch) for sale on the street. In the United Kingdom, heroin has a variety of names including (but not limited to) smack, horse, junk, Harry, and dragon. It is usually taken either by inhaling the heated fumes from a heroin mixture, a process known as "Chasing the Dragon", or by intravenous injection, a process known as "mainlining". Another process by which the drug is taken, although less frequently, is called insufflation, or "snorting" where the drug is ingested through the nose. Finally, the drug may be taken orally. Heroin samples may contain a mixture of different drug classes. For example, in the United Kingdom, street samples of

---

Controlled Drug Analysis
By Michael D. Cole, Lata Gautam and Agatha Grela
© Michael D. Cole, Lata Gautam and Agatha Grela 2025
Published by the Royal Society of Chemistry, www.rsc.org

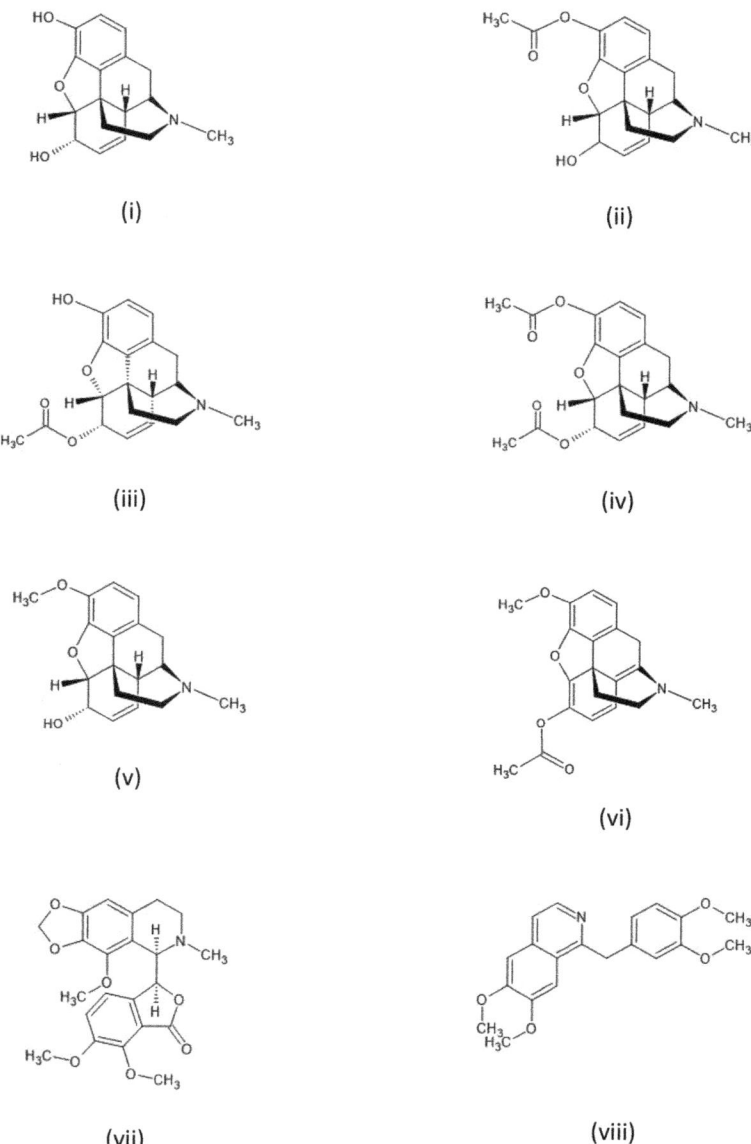

Figure 7.1 Examples of phenanthrene (i–vi) and benzoylisoquinoline (vii–viii) alkaloids found in heroin following the synthesis of diamorphine from morphine: (i) morphine, (ii) 3-O-monoacetylmorphine, (iii) 6-O-monoacetylmorphine, (iv) diamorphine, (v) codeine, (vi) 6-O-monoacetylcodeine, (vii) noscapine and (viii) papaverine.

heroin may contain diamorphine itself and other opiate alkaloids, barbiturates, which are present to prevent a pseudo-epileptic fit as the body clears the heroin and benzodiazepines, for example, diazepam or temazepam, as anxiolytics.

Heroin may also occur as part of a specific mixture of different classes of drugs. When drugs are taken together, this process is described as polydrug use. In the United Kingdom, an example includes taking heroin and cocaine together, a process known as "speedballing". Speedballing is also a generic term used to describe the process of taking a stimulant drug and a sedative drug together for recreational purposes. In the United States, heroin is frequently mixed with fentanyl or, more dangerously, carfentanyl. Both drugs are far more potent than the opiates themselves. Fentanyl is 100 times as potent as morphine, while carfentanyl is 10 000 times as potent as morphine. As a consequence, such mixtures present very real risks to drug users. In other parts of the world, heroin may be mixed with other types of drugs. In South Africa, for example, heroin may be mixed with methaqualone and smoked with tobacco. This mixture is known as "White Pipe". Alternatively, heroin may be mixed with tobacco, antiretroviral drugs such as lopinavir or efavirenz (which themselves are hallucinogenic), tramadol and cannabis in a drug mixture known as Nyaope, which is used in the townships of South Africa.[1] Drug chemists should be aware of the many classes of drugs with which heroin can be mixed.

In addition to the drug mixtures and seized ones themselves, which may range in size from single dose units to multiple kilograms, forensic scientists may also encounter paraphernalia associated with heroin use. This may include items where the heroin has been mixed on a surface, for example, an oven door or a mirror. Items may include scales that have been used to weigh the drug, and it can be recovered from the weighing surface. Other items may include syringes, spoons and other items associated with drug use. These are found where the heroin sample has been heated over a candle or a lighter flame and dissolved in water containing, for example, lemon juice in the bowl of a spoon. The former helps the drug dissolve in aqueous solution; the latter results in the salt form of the drug that is water soluble. The drug solution is drawn into a syringe through a cotton wool wad in an attempt to remove particulate material, from which the drug is then injected directly. In each of these cases, the item should be treated as a trace sample rather than a bulk sample and analysed accordingly.

Street samples of heroin may contain a wide range of concentrations of diamorphine. For example, some samples contain perhaps only 1 to 5% diamorphine, while other samples contain more than 90% diamorphine in the drug. The amount of drug that is taken depends upon the route of administration and the tolerance of the drug user. In terms of intravenous injection, someone with no tolerance may use between 5 and 15 mg of heroin. Someone with greater tolerance may use between 20 and 60 mg of heroin. Smokers may use 15 to 30 mg, and typically those who ingest heroin may take upwards of 50 to 70 mg.

In terms of legislation, opiate drugs are controlled by the United Kingdom Misuse of Drugs Act 1971. Diamorphine is a Class A drug and codeine a Class B drug. The impurities such as the 3- and 6-monoacetylmorphine are controlled as esters of morphine, and acetylcodeine is controlled as an ester of codeine. These drugs are controlled both in the freebase and in the salt forms. Heroin is an example of a drug that is strongly legislated for across the globe.

7.2 The Origins of Heroin

There are four major centres of production of heroin on a global scale. The first of these is found in what is known as the Golden Triangle – a region that is found overlapping Myanmar, Laos and Thailand. Another region where heroin is produced is known as the Golden Crescent and is found in the region overlapping Iran, Afghanistan and Pakistan. Interestingly, transit routes from this region of production are developing across the Indian Ocean and the east coast of Africa, where the heroin is landed and shipped across continental Africa towards the Mediterranean and into Europe. Such transit routes are becoming preferred because of the more porous borders and less stringent security offered on this route when compared to the Balkan route through which traditional heroin transit was made. To a lesser extent, heroin is produced in the Middle East, for example, in Lebanon and Syria. Heroin is also produced in Central and Latin America, including Mexico, Colombia, Peru and Bolivia.

7.3 The Manufacture of Heroin

Heroin is manufactured from morphine, which is found in the latex of the field poppy, *P. somniferum*. The field poppy is the only poppy to produce morphine in significant quantities. In order to collect the opium resin from which morphine is extracted, the opium farmers allow the poppies to flower. Once the seed head has developed, it is incised after the petals have fallen off and the resulting white sap is allowed to exude from the seed head. It is the dried sap which is known as opium which contains between 4 and 25% morphine by weight. It also contains codeine, noscapine (which is also known as narcotine), papaverine and thebaine (Figure 7.1). The opium resin itself can be smoked, eaten or turned into a drink. However, in order to manufacture diamorphine, further processing of the opium is required. Due to the chemical characteristics of the alkaloids in addition to morphine that are found in opium, they carry through from the raw opium into the final product of heroin. This is because their chemical structures, pK_a values and solubilities in acids, bases and organic solvents are not dissimilar to diamorphine itself. It is these alkaloids that provide the forensic scientist with the markers to determine whether a sample is illicitly synthesised heroin or from a medical source, the latter of which is chemically pure, and also enable the comparison of heroin samples. One route by which diamorphine is synthesised is by boiling opium with a mixture of calcium hydroxide, ammonium chloride, ethanol and diethyl ether in water. The opium dissolves and as the composition of the solvent changes, due to the different rates of evaporation of the different solvents in the mixture, the crude morphine precipitates out of the mixture. This is then dissolved in sulfuric acid and refluxed in the presence of charcoal. The sulfuric acid turns the alkaloids into the sulfate salts, rendering them water soluble. Charcoal is used to remove the low-molecular-weight plant phenols and decolourises the mixture. Following this process, the sample is basified, commonly with ammonium hydroxide, and morphine precipitates from the mixture, this time as a much cleaner product. The morphine is then reacted with acetic anhydride, which acetylates the morphine (and similarly the hydroxide group of codeine), resulting in the synthesis of diamorphine (and similarly, acetylcodeine). The diamorphine is then further purified using a series of acid–base extractions. In terms of yield, 10 kg of morphine will result in approximately 1 kg of pure diamorphine.

As a consequence of the differences in the botany of the poppy plants, how they are grown, the environmental conditions pertaining to where the plants are grown and agronomical differences, along with differences in the synthetic routes used, all of which result in different amounts of the original opiates in the opium resin, reaction intermediates and final products, there are distinct chemical differences between heroin batches. These differences can be used to address a number of questions relating to heroin, including

1. Is the sample heroin and what/how much drug(s) does it/do they contain?
2. Are samples of heroin related to one another?
3. Where do the samples come from?

The techniques described below allow the forensic drug chemist to address these questions.

7.4 Physical Description of Heroin Samples

A wide variety of materials containing heroin may be encountered in the forensic chemistry laboratory. Materials range from opium resin through brown powders or granular material to highly refined powdered white drug. Under such circumstances, a full physical description of the materials is required, including an examination of whether or not resinous materials fit together. The description should include details of the labelling of the item, the colour, weight and smell of the material. The smell is important because heroin samples typically smell of acetic acid, a by-product of the synthetic process of the acetylation of morphine. In addition to the description of the drug material, any wrapping material should be fully described, including whether or not the wrapping materials fit together or with items seized from other locations – for example, cling film rolls seized from the house of those suspected of distributing heroin. Such physical resemblances instantly establish a link between samples. Powdered and resinous materials should then be sampled according to the protocols described in Chapter 2.

A wide variety of paraphernalia associated with the distribution and use of heroin may also be encountered. Such samples include surfaces upon which heroin samples were mixed, for example, mirrors or oven doors, spoons and syringes used in the preparation of heroin for injection, scales used to weigh out heroin samples, and in some countries, the vessels in which heroin samples have been prepared. These items should be treated as trace samples. A fraction of the material to be analysed on the surfaces to establish whether or not heroin is present should be removed using a swab soaked in solvent. The reason that the whole of the material is not removed is to allow for a second and subsequent analysis, which is especially important in adversarial judicial systems. The solvent used should not react with the drugs being analysed (in this case, opiates) and ideally should be compatible with subsequent analyses. This is why chloroform is frequently chosen. The samples should then be extracted into the same solvent, any solid material removed, usually by centrifugation, which reduces sample loss and the risk of contamination and the extract should then be subject to instrumental analysis.

7.5 Presumptive Tests for Heroin Samples

Heroin samples, by their very nature, are often complex mixtures of a variety of different drug classes. However, the principal presumptive test for heroin is the Marquis reagent (5 ml of 40% formaldehyde in 100 ml of concentrated sulfuric acid). When placed on a heroin sample, this will give a deep blue/purple colour reaction. However, a blue reaction (but different from the product formed with opiates, which has a different colour) is also obtained from ring-substituted amphetamines. Consequently, both positive and negative controls should always be completed too. In addition to the opiate drugs, heroin samples may contain barbiturates, benzodiazepines, methaqualone, cocaine, tramadol, fentanyl and related compounds, and amphetamines in addition to a wide variety of adulterants and diluents including caffeine and paracetamol (acetaminophen). As a consequence a wide variety of tests are required for the determination of drug classes thought to be contained in heroin samples, as described throughout this book. On the basis of the results of the presumptive tests, TLC can be used to determine which members of a drug class are present in the heroin sample, and following consideration of these results, the instrumental method of analysis can be chosen.

7.6 Thin Layer Chromatography of Heroin

Given the complex nature of heroin samples, TLC is a useful technique for screening for which opiates are present and also for comparison of heroin samples. There are a number of TLC systems described for the analysis of heroin.[2] However, the first step in TLC analysis is the dissolution of the sample in a suitable solvent. For heroin, chloroform is ideal because it is a solvent in which all of the opiate drugs are stable and freely soluble. Samples can be dissolved at a concentration of 1 mg street drug per ml because this will give an opiate concentration suitable for application to the origin of the thin layer chromatogram and comparison with standard drugs. The samples should be analysed alongside opiate standards and have been freshly prepared again at a concentration of 1 mg ml^{-1} in chloroform. Once the samples have been applied to the TLC plate, which is usually silica gel impregnated with a fluorescent indicator, they should be allowed to dry and the chromatogram developed in one of a number of different mobile phases. These include

1. toluene, acetone, ethanol and concentrated ammonia (45 : 45 : 7 : 3 by volume);
2. ethyl acetate, methanol, and concentrated ammonia (85 : 10 : 5 by volume); and
3. methanol and concentrated ammonia (100 : 1.5 by volume).

These mobile phases are all suitable for the analysis of samples containing opiates, and each contains ammonia. The ammonia ensures that the drug is in the free base form (all opiates are bases) and, as a consequence, exhibits good chromatographic properties and does not tail. Once the chromatogram has developed, it should be removed from the TLC tank, the solvent front should be recorded, and the chromatogram should be allowed to dry. After drying, the chromatogram should be observed under white light, ultraviolet light at 254 nm, where any compound present will appear as purple spots on a yellow background, and at 360 nm where some compounds may fluoresce. At each stage, any compound observed should be recorded. The chromatogram can then be visualised using either acidified potassium iodoplatinate (0.15% potassium chloroplatinate and 3% potassium iodide in dilute hydrochloric acid) or Dragendorff reagent (commonly 0.85 g of bismuth subnitrate, 40 ml of water, and 10 ml of glacial acetic acid mixed with an equal volume of 8 g of potassium iodide dissolved in 20 ml of water and then diluted with water and glacial acetic acid). The former will give a variety of purple and blue colours with the opiates and other compounds present, while the latter gives an orange reaction with alkaloidal drugs. Again, any compound present should be recorded. Where compounds are observed, this gives an indication of standards that should be used for instrumental analysis. The profiles on the thin-layer chromatogram can also be used as a quick screening method for the comparison of drug samples, and using this technique, rapid exclusions where there are questions about similarity can be made. Having completed the TLC analysis, the analyst should move on to instrumental methods.

7.7 High-performance Liquid Chromatography of Heroin

A large number of liquid chromatography systems have been described for the analysis of opiates in heroin samples.[3] Prior to analysis, the samples should be prepared in a suitable solvent. Ideally, this is the mobile phase of the HPLC system itself, but regardless of the solvent used, it must be completely miscible with the HPLC mobile phase. Any particulate material should be removed by filtration or, preferably, centrifugation. The separations are generally effected on reversed-phase C-18 columns, with the separation mechanisms being a mixture of partition, ion exchange and size exclusion. The method chosen depends on the equipment available in the laboratory. Normal-phase chromatography has also been applied to the phenanthrenes in heroin samples,[4] using a silica gel column (12.5 cm × 4.6 mm i.d.), with a mobile phase of iso-octane/diethyl ether/methanol/water/diethylamine (400 : 325 : 225 : 15 : 0.65 by volume), with a flow rate of 2 ml min^{-1}, with detection at 230 nm. In addition to providing a relatively rapid analysis time compared to other methods (<10 min), a number of chromatographic principles can also be illustrated with this system. The diethylamine renders the mobile phase basic, suppressing the ionisation of the drugs, which are basic with pK_a values ranging from 7.95 to 10.19. The separation mechanism is principally sorption/desorption and is based upon the strength of the interaction of the analytes with the silica gel stationary phase (Table 7.1). The system effects a separation based primarily upon the polarity of the analytes, with the least polar diamorphine eluted first and morphine last. Usefully, noscapine and papaverine are also separated using this system (Figure 7.2).

A variety of detection techniques are also possible. The simplest of these uses UV detection, which allows drug identification on the basis of retention times and at the same time offers the required linear dynamic range and sensitivity to quantitate the drugs should this be necessary.

If dual-wavelength detectors (monitoring at two wavelengths of UV light between 210 and 350 nm) are used, it is possible to add weight

Table 7.1 Nature of sorption/desorption interactions of morphine and diamorphine with silica gel, which result in their separation.

	Morphine	Diamorphine
Pi electron interactions (benzene rings and double bonds)	Yes	Yes
Ether oxygen lone pair	Yes	Yes
Hydrogen bonding with a phenolic group	Yes	No
Hydrogen bonding with enolic alcohol	Yes	No
Hydrogen bonding with carbonyl groups	No	Yes
Hydrogen bonding with a nitrogen atom lone pair	Yes	Yes

to the identification. Based on the absorbance of each opiate at the chosen wavelengths, it is possible to calculate absorbance ratios to (i) identify the eluted compound because the ratio will be different for each analyte and (ii) determine whether there is any co-elution with the opiate. This is possible because a pure compound will provide a near square wave as the absorbance ratio is monitored over time, while if there is co-elution, assuming that the co-eluting molecule also absorbs at the relevant wavelength, the output will not be a square wave but some other shape.

Further evidence for compound identity is achieved using diode array detection where the absorbance maxima, minima and shoulders provide evidence of the compound identity in addition to the retention time. For the latter technique, it is also essential that the mobile phase used is from the same batch for the standard compounds and experimental samples used in the analysis as variations in pH will affect the absorbance maxima measured.

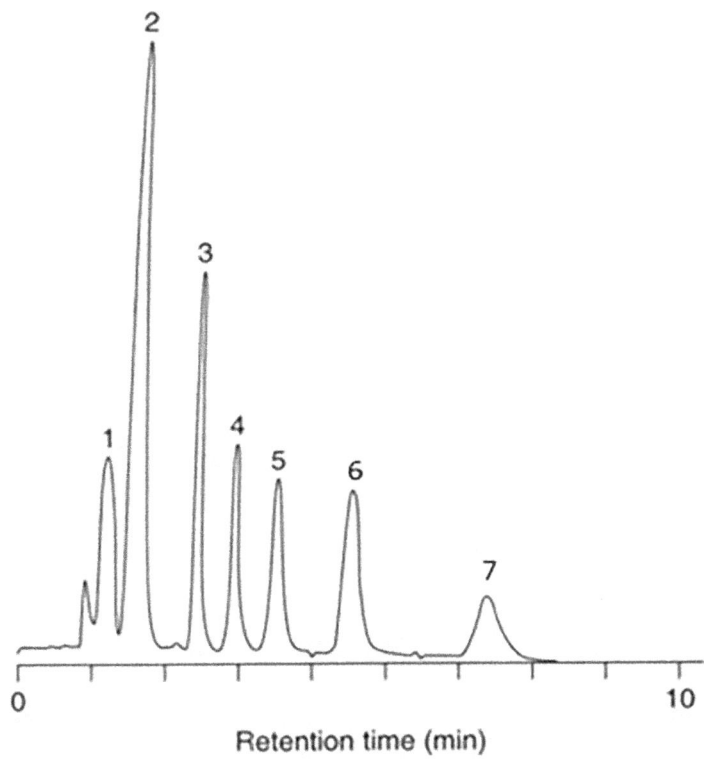

Figure 7.2 Separation of compounds commonly found in heroin by normal-phase HPLC. 1 = noscapine, 2 = papaverine, 3 = diamorphine, 4 = acetylcodeine, 5 = 6-O-monoacetylmorphine, 6 = codeine and 7 = morphine.

7.8 Gas Chromatography–Mass Spectrometry of Heroin

Across the globe, GC–MS is the method of choice for the analysis of heroin samples. The opiates and materials added to heroin can be analysed in derivatised or underivatised forms. This may be necessary when, for example, other drug classes are present, including cannabinoids, methaqualone or cocaine. The hydroxide groups present on morphine, codeine and monoacetylmorphine will cause tailing if the instrument, especially the chromatographic column, is not scrupulously clean, but with proper instrumental maintenance, this should not be a problem.

Representative samples of the material to be analysed should be taken (Chapter 2) and fully homogenised. Solutions should then be prepared for analysis. Chloroform is frequently used as the opiates are stable in this solvent, it is volatile and it does not react with the analytes under consideration. In general, low-molecular-weight alcohols (*e.g.* methanol) should be avoided because they contain water and will result in analyte hydrolysis and artefact formation during the GC–MS analysis. However, some drugs mixed with heroin, for example, cannabis, are not stable in chloroform. Under such circumstances, an anhydrous tertiary alcohol may be used, as when Nyaope is analysed.[1] Under other circumstances, it may be necessary to experiment with solvents, but the general criteria that the solvent should be volatile, not react with the drugs being analysed, should not catalyse their decomposition and be a solvent in which they are freely soluble over a wide range of concentrations should be adhered to. Solutions of standard drugs and blanks of solvent alone should also be prepared for analysis. Once the sample is dissolved, particulate matter that may be present should be removed by either filtration or centrifugation. It might be argued that the latter is preferable to the former as there is always the risk of contamination of the sample by material on the filter or loss of material by sorption onto the filter surface. The sample can then be analysed directly, following the addition of an internal standard, or following derivatisation.

If derivatisation is employed, the reagent should "protect" all of the hydroxide groups present on the opiate drugs. A number of reagents can be employed, including *N,O*-bis(trimethylsilyl)acetamide or acid anhydrides (Figure 7.3). Acetic anhydride should be avoided as a derivatising reagent because it will convert morphine, monoacetylmorphine and codeine into acetylated derivatives, which would also be found in the heroin sample. To derivatise samples, an aliquot of the solution to be analysed is placed in a GC and blown down under dry nitrogen. A fixed volume of reagent, also containing internal standard, is added to each vial. The sample may then be allowed to derivatise at room temperature, or with heating, as appropriate to the reagent. Following derivatisation, the samples, controls and blanks are analysed in the order blank, derivatised blank, derivatised standard, derivatised blank, derivatised sample, derivatised blank *etc*. The analogous sequence is used for underivatised samples. Using this sequence, it is demonstrated that the reagents and instruments are clean, that the derivatisation (where used) works, that retention times and reference mass spectra under the experimental conditions are obtained, that there is no "carry over" between samples and that the drugs in the samples can be identified.

Typically, the following instrumental parameters can be used for underivatised or derivatised samples. The column may be OV-1, BP-1, BP-5 or equivalents, 25 m × 0.22 mm i.d. × 0.5 μm. An injection temperature of 250 °C, with a column oven temperature programme of 150 °C for 2 min, increasing to 280 °C at 9 °C min^{-1}, with a 2 min hold is typically used. The carrier gas used is helium with a typical flow rate of 1 ml min^{-1}. The split ratio will depend on the sample concentration but can be as large as 50 : 1. Mass spectrometric detection is usually employed with a filament voltage of 70 eV.

Typically, as TMS derivatives, the compounds elute in the order: TMS derivative of codeine, acetylcodeine, morphine, 6-O-monoacetylmorphine and diamorphine. The elution order is a function of boiling point and derivative. Compounds are identified on the basis of retention time or, better, retention index relative to an internal standard, and mass spectra. An example of the data generated is shown in Figures 7.4 and 7.5.

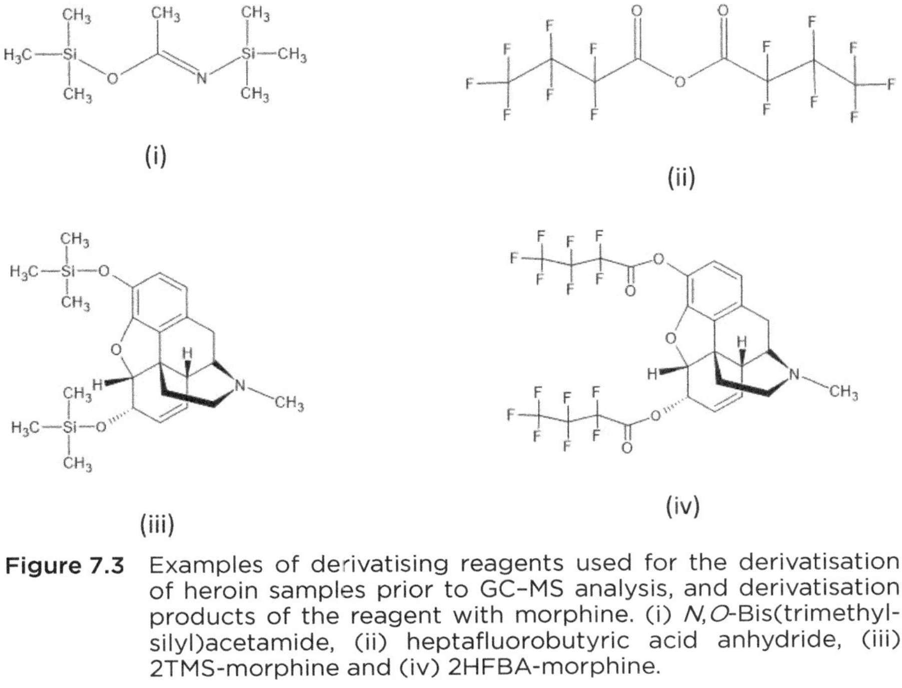

Figure 7.3 Examples of derivatising reagents used for the derivatisation of heroin samples prior to GC–MS analysis, and derivatisation products of the reagent with morphine. (i) N,O-Bis(trimethylsilyl)acetamide, (ii) heptafluorobutyric acid anhydride, (iii) 2TMS-morphine and (iv) 2HFBA-morphine.

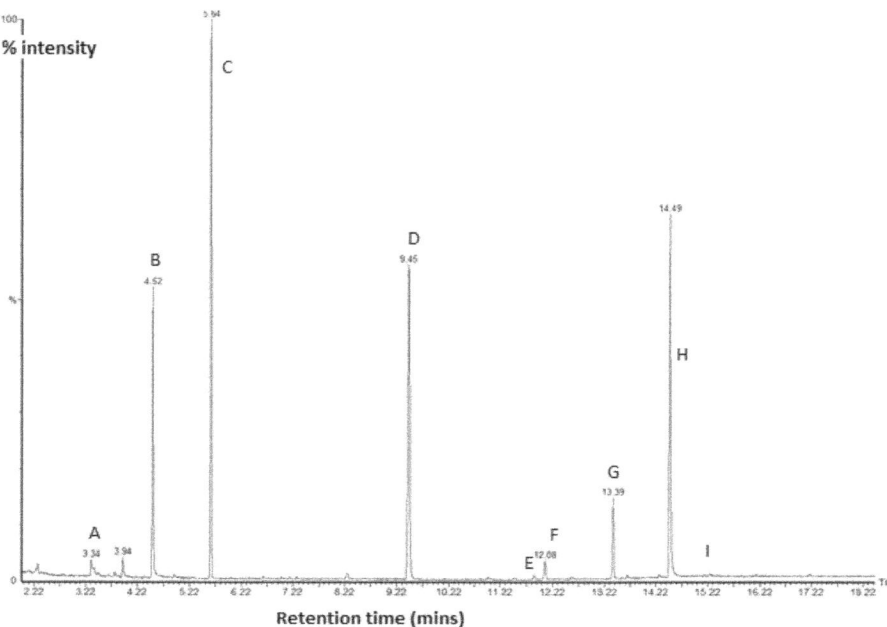

Figure 7.4 Chromatogram of the heroin street sample. (A) Acetaminophen, (B) caffeine, (C) eicosane (internal standard), (D) tetracosane (internal standard), (E) acetylcodeine, (F) 6-monoacetylmorphine (6-MAM), (G) diacetylmorphine, (H) octacosane (internal standard), and (I) papaverine. Retention index of diamorphine = 0.92. The retention index for the standard diamorphine against octacosane was also 0.92.

7.9 NMR Spectroscopy of Heroin Samples

It is widely accepted that GC–MS is the gold standard method for the analysis of heroin, allowing both identification and quantification of the components of a heroin mixture. The technique does, however, require that the analytes are thermally stable and are fully volatile at temperatures below 350 °C. It is also known that some drugs found to be mixed with heroin do not satisfy these criteria. These include some of the benzodiazepines and some antiretroviral drugs found in polydrug mixtures such as Nyaope. Under such circumstances, another analysis technique may be required. It is now becoming recognised that ^{1}H NMR is a powerful tool that can be used in forensic science, both to identify components of complex mixtures and to quantify them. It also has the advantage that the whole sample can be presented for analysis without sample purification,

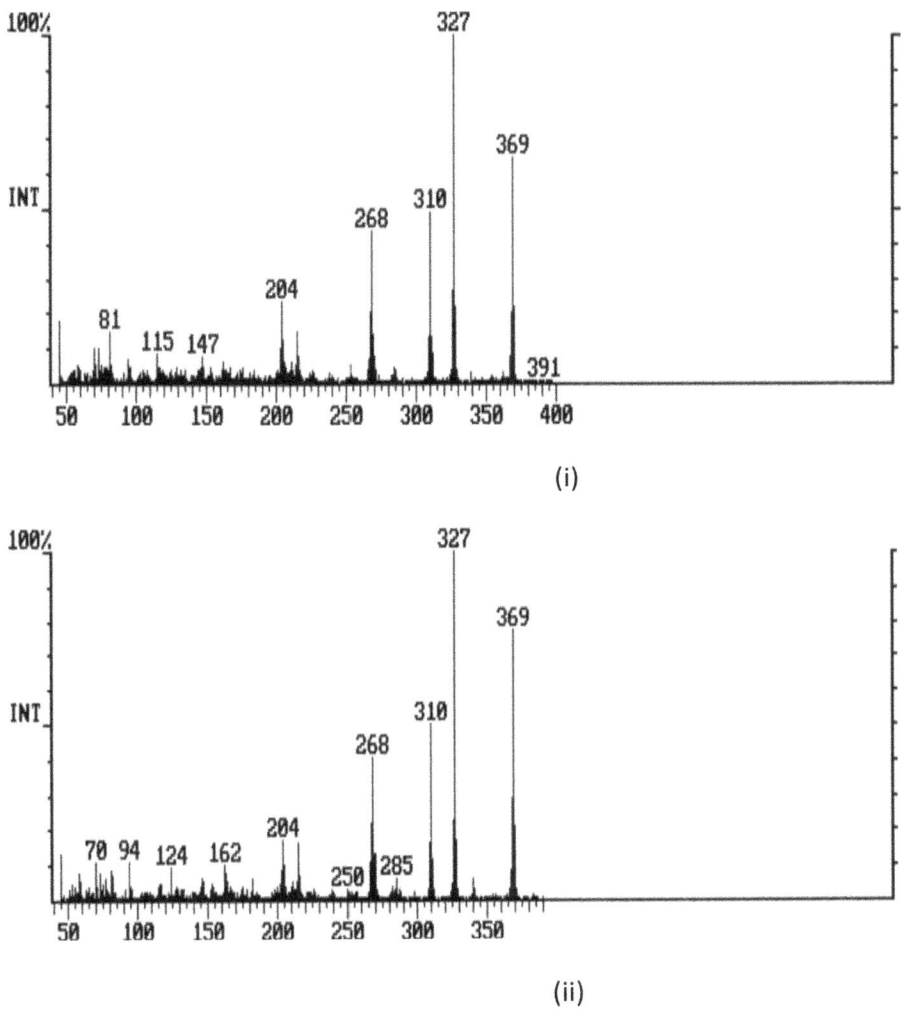

Figure 7.5 Mass spectrum of diamorphine following GC–MS analysis of (i) diamorphine standard and (ii) diamorphine in a street sample of heroin.

separation or derivatisation. NMR has the further advantage that samples can be recovered after analysis.

NMR therefore provides another technique that may be useful in heroin analysis, which has been proven to be the case.[5] Milligram amounts of the sample can be dissolved in d_6-DMSO. DMSO is a good solvent because the wide range of compounds found in heroin are freely soluble in the solvent, with the deuteration providing the lock for the NMR. Ahead of experiments, the samples, after dissolution, can be centrifuged to remove any solid material that will interfere with the quality of the spectra obtained. With the addition of a suitable internal standard, it is possible to achieve accurate chemical shift measurements and quantify the drugs in the sample. In order to quantify the drugs, a signal at a given chemical shift that is unique to the molecule being analysed must be identified. The size of this signal can then be compared to another unique signal of an internal standard of known concentration.

NMR is also sufficiently sensitive to achieve the identification and quantification of diamorphine and related alkaloids in street samples. Limits of detection (LOD) have been described at around 3 µmol l^{-1} and limits of quantification (LOQ) around 20 µmol l^{-1}. These correspond to diamorphine concentrations of 0.02% and 0.15% by mass, respectively. The precision of NMR is also reported to be similar to GC–FID and GC–MS techniques, with measured precision lying between 2% and 7%.

In addition to allowing the identification of opiate alkaloids in heroin samples, as well as numerous adulterants and diluents across a wide range of concentrations, it is also possible to achieve chemical fingerprinting of heroin samples and compare them with each other. This can be achieved using standard ^{1}H NMR experiments in addition to an NMR technique called two-dimensional diffusion-ordered spectroscopy.[5] As such, NMR provides a powerful tool for the identification and comparison of heroin samples, but being expensive, it does not find widespread application in forensic science laboratories in routine casework. This situation may change with the development of benchtop techniques.

7.10 The Analysis of Stable Isotopes in Heroin

Stable isotopes offer an opportunity to not only compare heroin samples but also determine their geographic origin. Not only the isotopes of carbon and nitrogen have been used to determine origin[6,7] but strontium has also been used.[8] This is a particular advantage of stable isotope analysis over chromatographic methods, which have been primarily used for the identification, quantification and comparison of heroin samples. The method works because the stable isotope ratios are determined as the opiate alkaloids are synthesised in the plant, particularly the isotopes in the morphine component. The ratios may be changed slightly because the morphine is extracted from the plant material and then acetylated. The method also requires reference samples from the various heroin production areas of Southwest Asia, Southeast Asia, Mexico and South America so that reference ratios may be determined. The ratios are expressed as δ, units of ‰, and calculated as

$$\delta\% = (R_{sample}/R_{standard} - 1) \times 1000 \tag{7.1}$$

where the standards are certified reference materials. Work has shown that the isotope ratios differ by values of 2.4‰ and 3.1‰ for carbon and nitrogen, respectively, and that it was possible to discriminate between heroin from the four regions where it is manufactured. When the diamorphine is hydrolysed to give the original biosynthetic product, morphine, the method becomes even more discriminating because the variation added by the different batches of acetic anhydride, used to produce the diamorphine, is removed, leaving only the core morphine, the product of biosynthesis in the plant.

The drug chemist should also consider the number of batches or "crops" that have been obtained during the manufacturing process. For example, it has been shown[8] that nitrogen isotope ratios do not vary between the first and second batches of heroin from a particular synthesis, but the isotope ratios in the third batch (and presumably subsequent batches) can be significantly different. This is because those molecules containing the greatest proportion of heavier isotopes precipitate ahead of those composed of the lighter isotopes. It is for this reason that very great care should be taken in the interpretation of stable isotope ratios in heroin, and the ratios associated with the native morphine will give the greatest opportunity for the accurate determination of the geographic origin of the sample.

In addition to the analysis of carbon and nitrogen isotopes, $^{87}Sr/^{86}Sr$ ratios have been examined as potential markers for the geographic origin of heroin.[8] The use of isotopes such as these requires knowledge of the ratios in the bedrock, soil and groundwater in which the plants, in this case *Papaver somniferum*, are growing and make the assumption that the ratios in these sources of minerals are the same as those in the growing plants. Using the strontium ratios, it was possible to correctly identify the origin of the heroin up to just above 80% of the time. What would be interesting is the combination of carbon and hydrogen in the native morphine, coupled to the isotopic ratio analysis of other trace elements to determine whether further discrimination of the geographic origin of the heroin is possible.

7.11 The Analysis of Metal Ions in Heroin

While the presence of metals cannot be used alone to identify heroin, the analysis of metal ions can be used to compare heroin samples. The inorganic ion content arises from a number of different sources. These include inorganic ions being contained within the opium resin itself, which is harvested from the plants, being dissolved in the solvents and reagents used to prepare the heroin from the opium resin and leachates from the vessels and utensils in which the heroin is prepared. Such analyses provide an opportunity for multivariate analysis to be used to classify (in terms of origin) and cluster (group together) heroin samples.

The vast majority of methods of inorganic analysis of heroin involve the digestion of the inorganic content of the heroin in nitric acid (all metal nitrates are soluble in water) and hydrogen peroxide, with subsequent analysis by, in general, ICP-MS. Polyethylene vessels are used for the digestion and analytical processes because they are less likely to leach inorganics into the sample being analysed. The drugs are then quantitatively analysed for a number of inorganic elements using principal component analysis, partial least squares discriminant analysis and hierarchical clustering analysis frequently used as tools, applied to the data, from which to draw conclusions. In order to develop meaningful clusters from such data, the assumptions underlying such chemometric methods should be tested to ensure that they are met before any conclusions are drawn.[9]

7.12 The Analysis of Occluded Solvents in Heroin

In addition to the analysis of the drug components of a heroin sample, it is possible to analyse the solvents that may be contained in the crystalline structure of the drug in order to compare samples and determine whether or not they may be related to each other. Such analyses do not provide evidence of the identity of a heroin sample but can be used as part of a suite of evidence to establish a relationship between materials. Heroin is made by the extraction of morphine from opium, subsequent acetylation and then purification using a series of acid–base extractions and crystallisations (Section 7.3). Under such conditions, the solvents that are used in the chemical reactions and to recover the subsequent products may become enclosed in the drug crystals. These are known as occluded solvents and may include starting materials, reaction intermediates, and solvents used during the extraction processes.

Straightforward techniques taken from technologies used to recover ignitable liquid residues from fire debris can be used to recover occluded solvents from the headspace above a drug sample.[10,11] If such analyses are to be carried out, they should be performed as soon as possible after the seizing of the drugs. This is because through evaporative processes, the occluded solvents present in the sample will change over time. Simple experiments can be used to recover headspace samples. Between 250 and 300 mg of drug samples can be placed in a small vial. The headspace above the sample should be small so that the vapours are sufficiently concentrated that they can be recovered. In addition to the drug, a small charcoal strip should be inserted into the vial inside, for example, a glass insert. This allows the charcoal strip to sorb the vapours from the headspace but at the same time not be in direct contact with the drug sample. The sample and charcoal strip can be heated, generally to 80 °C for 60 minutes. Following this, the charcoal strip can be recovered and the absorbed solvents eluted using a small amount of a suitable solvent, for example, carbon disulfide. Carbon disulfide is preferred to pentane or hexane because it is extremely unlikely to occur as a solvent that has been used in drug preparation, unlike the latter solvents. Additionally, it is unlikely to react with the vast majority of common solvents. The eluted compounds can be analysed by gas chromatography with either flame ionisation detection or mass spectrometric detection. The solvents can be identified on the basis of retention time and/or mass spectrum and the comparison made between the chromatographic profiles obtained from the different heroin samples.

Using these techniques, some solvents are found to be quite common in heroin samples, while others are less so. It is the less-frequently encountered solvents that provide greater weight of evidence. Care should be taken in interpretation because the relative standard deviations of sample concentrations in some studies range from 80% to 620%. The interpretation of this is that if the included solvent profiles match, then the samples once formed a larger batch.

If they do not match, it is more difficult to interpret the data because either the samples are not at all related or they are indeed related but that the occluded solvents have evaporated at different rates depending on the environment in which they have been kept and stored.

Care should also be taken around where the solvents have come from within the heroin sample.[12] For example, some adulterants, including sugars and caffeine, do not affect the occluded solvent profile of the drug. However, some adulterants added to heroin, for example, local anaesthetics, including lidocaine, have been found to affect the occluded solvent profile because the adulterants themselves have included solvents that are recovered in addition to those from the heroin.

A further complication is that it is possible that the occluded solvents themselves break down during drug storage and the profile of occluded solvents is in fact artefactual. For example, it is possible that ethyl acetate under acidic conditions will hydrolyse to ethanol and acetic acid. It is the ethanol and acetic acid which are found during the analysis. Whether or not this is actually occurring can be determined using deuterated d_8-ethyl acetate, which if hydrolysis occurs will lead to deuterated ethanol and acetic acid being observed. Finally, artefactual production of chlorinated solvents may occur. For example, chloroethane has been found in drug samples because of the bubbling of hydrochloric acid gas through solutions of the drug in order to manufacture the hydrochloride salt. The hydrochloric acid reacts with the solvents forming chlorinated counterparts.

Where occluded solvents have been used in heroin comparisons, weighting should be used in terms of the interpretation and strength of the evidence involved. It is generally accepted that the profile of the drug impurities provides the primary evidence regarding whether two or more drug samples are linked. Occluded solvents provide supporting evidence for drug linkage when both the organic profile and the occluded solvent profile match. Where the organic drug profile matches but the occluded solvent profile does not, the interpretation is rather more difficult, and consideration should be given to whether or not the occluded solvent profile may have changed because of storage conditions.

7.13 DNA Analysis and Heroin

There are a large number of reasons why the opiate and inorganic content of heroin may vary. These include variations in how the field poppies are grown, the soil, climatic and environmental conditions in which they are grown and differences between cultivars. In addition, there may be variations in the way that opium and morphine are processed, both between regions and with time. This may lead to difficulties in using organic or inorganic impurities to compare heroin samples and relate them to each other. Under such circumstances, it is necessary to find another mechanism by which heroin might be identified and samples compared. Included in the toolbox of the forensic scientist is DNA analysis. Despite the considerable processing of opium and morphine during the manufacture of heroin, it has proven possible to extract DNA from different heroin types of heroin samples, ranging from black tar through to highly purified powder, and identify the source of that material as the field poppy, *P. somniferum*.[13] This is despite the extremely strong alkalis, acids and high temperatures that are used during opium processing.

Heroin samples vary in the quantity of diamorphine and other opiate alkaloids in the samples as well as the quantity and nature of the adulterants and deliverance that are found in the materials being analysed. This has a direct bearing on the quality and quantity of DNA that can be extracted from the samples. In order that DNA can be extracted and analysed from heroin, a relatively large sample is required. Current work requires that hundreds of milligrams to gram amounts of sample are available to the analyst. Using commercially available DNA extraction kits, it has proven possible to extract enough DNA to carry out PCR using three DNA markers specific to the field poppy and subsequently carry out capillary electrophoresis analysis of the DNA from the samples. While this is an extremely useful technique, it does not currently find wide application in forensic science laboratories, and studies are needed to determine whether heroin samples can be both clustered and classified according to their origins.

References

1. P. M. Mthembi, E. M. Mwenesongole and M. D. Cole, *Emerging Trends in Drugs Addict. Health*, 2024, 100142.

2. United Nations, in *Recommended methods for testing opium, morphine and heroin*, United Nations Office on Drugs and Crime, New York, 1998.
3. K. Diekhans and I. S. Lurie, *Forensic Chem.*, 2022, **31**, 100455.
4. P. C. White, *personal communication*.
5. S. Balayssac, E. Retailleau, G. Bertrand, M. P. Escot, R. Martino, M. Malet-Martino and V. Gilard, *Forensic Sci. Int.*, 2014, **234**, 29.
6. J. R. Erhleringer, D. A. Cooper, M. J. Lott and C. S. Cook, *Forensic Sci. Int.*, 1999, **106**(1), 27.
7. J. F. Casale, J. R. Ehleringer, D. R. Morello and M. J. Lott, *J. Forensic Sci.*, 2005, **50**(6), JFS 2005077-7.
8. J. DeBord, A. Pourmand, S. C. Jantzi, S. Panicker and J. Almirall, *Inorg. Chim. Acta*, 2017, **468**, 294.
9. S. Ratshonka, Available from: https://enfsi.eu/wp-content/uploads/2022/11/EAFS-Abstract_web_flipbook.pdf. Accessed 25 Jun 2024. p. 438.
10. J. Cartier, O. Gueniat and M. D. Cole, *Sci. Justice*, 1997, **37**(3), 175.
11. M. D. Cole, *Forensic Sci. Rev.*, 1998, **10**(2), 113.
12. J. R. Mallette and J. F. Casale, *J. Forensic Sci.*, 2015, **60**(1), 45.
13. M. A. Marciano, S. X. Panicker, G. D. Liddil, D. Lindgren and K. S. Sweder, *Sci. Rep.*, 2018, **8**(1), 2590.

8 The Analysis of Amphetamines, Ring-substituted Amphetamines and Related Compounds

8.1 Introduction

Amphetamines, of which there is now a wide variety in the drug market, are a group of drugs that act as stimulants. The ring-substituted variants of this drug group, such as the 3,4-methylenedioxy compounds, are also entactogens and induce a feeling of well-being. Their mode of action is *via* the effects that they have on the central nervous system. While there are many of these drugs in the market, the most commonly encountered of these are amphetamine, methamphetamine, 3,4-methylenedioxyamphetamine (MDA), 3,4-methylenedioxymethamphetamine (MDMA), and 3,4-methylenedioxyethylamphetamine (MDEA) (Figure 8.1). All are based on the basic amphetamine (β-phenethylamine) structure. There are a wide variety of structural variants, stereoisomers and regioisomers described both in the printed literature[1] and in internet sources. In PIHKAL[1] alone, 179 different compounds are described. These include different ring substitutions of the benzene ring, positional isomers involving the methylenedioxy group, and different substituents on the nitrogen atom. The vast majority of these were synthesised to (i) achieve a different type of effect and (ii) bypass legislative instruments introduced to control amphetamine-type drugs. The general principles applied to the amphetamines described in this chapter apply equally to these variants.

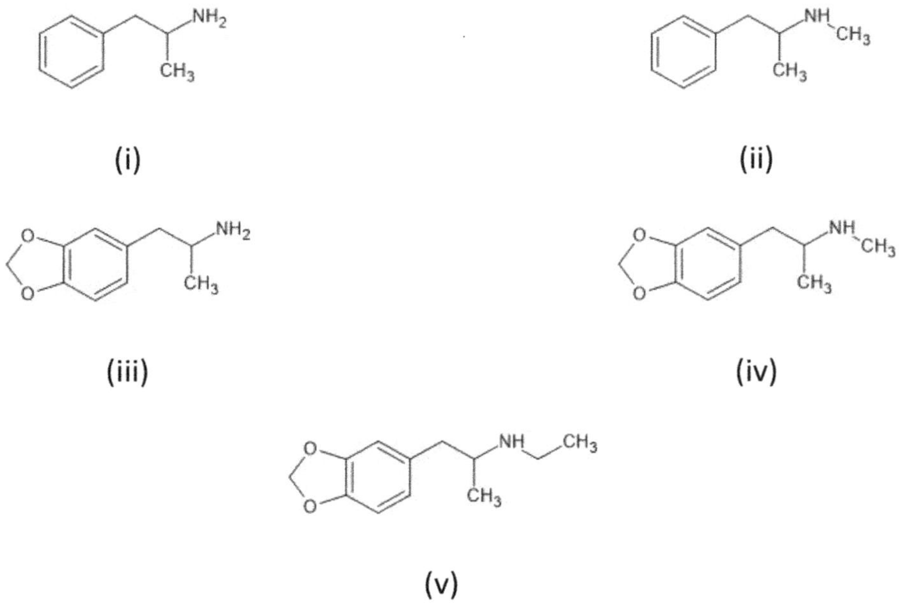

(i)

(ii)

(iii)

(iv)

(v)

Figure 8.1 Structure of (i) amphetamine, (ii) methamphetamine, (iii) 3,4-methylenedioxymethamphetamine (MDA), (iv) 3,4-methylenedioxymethamphetamine (MDMA) and (v) 3,4-methylenedioxyethylamphetamine (MDEA).

Street samples of these drugs include powders, which may be white to off-white, pinkish grey, light brown oils (which are usually the free base of the drug) or tablets. When in tablet form, these drugs can be taken orally. Alternatively, in powdered form, the drugs can be crushed and snorted (insufflated) or taken orally by placing the powder either sublingually or dabbing it onto the gums. Some users wrap the drugs in paper and swallow them – a process known as "bombing". Others smoke the drug either in a cigarette or in a specially designed apparatus. This is particularly true for methamphetamine. The drugs may also be injected or mixed with drinks. The dose of the drug depends on how much and how frequently the user takes it and the desired effect. The onset time (time taken for the drug to take effect) depends on the route of administration, with, in general, intravenous injection being the quickest and oral ingestion the slowest. The effect of the drug and its duration depends again on the user, the route of administration, and the drug itself. Exemplar data are given in Table 8.1.

These drugs are either totally synthetic, or semi-synthetic, that is derived from compounds from plants. The synthesis of these drugs

Table 8.1 Typical doses of amphetamine-type drugs used by recreational users, ranges of onset times and duration of action.

Drug	Usual dose (mg)	Onset time (min)	Duration of effect (h)
Amphetamine	5–50	0–60	3–6
Methamphetamine	5–150	0–70	3–5
MDA	30–300	30–90	3–6
MDMA	30–200	20–70	3–5
MDEA	30–200	—	2–4

leaves "tell-tale" traces of drug impurities within the drug sample, which arise from the starting materials, incomplete reactions and side reactions related to the synthesis. It is these impurities that allow the drug chemist to go beyond identifying and quantifying the drug, enabling them to both identify the routes by which the drug was synthesised and assess whether or not drug samples are related to each other. In addition, a number of catalysts may be used in the synthetic process. These catalysts themselves leave behind traces of the metal in the sample. It is the chemical signature of the metals that can also be used, sometimes, to determine the synthetic route and/or to compare drug samples. Further, the precursor chemicals are manufactured in batch-wise processes. It is these differences that mean that different batches will have different stable isotope ratio contents, for example, $^{12}C/^{13}C$ and $^{14}N/^{15}N$ ratios, and these differences and similarities can be used to confirm or exclude the relatedness of different drug batches.

The drugs themselves may contain a wide variety of adulterants and diluents. The adulterants are pharmacologically active and include substances such as caffeine, which itself is a stimulant, that are mixed with the sample to hide the absence of any of the amphetamine drugs. Diluents include materials like starch, which simply bulk the drugs out, increasing the profit made from selling the drugs, and materials such as stearates, which are used to facilitate powder flow during the course of pressing tablets during tablet manufacture.

8.2 Synthetic Routes for Amphetamine Synthesis

There are a wide variety of synthetic routes published for the synthesis of amphetamine, as exemplified in PIHKAL[1] and on the internet. Analogous routes are available for methamphetamine, further N-alkylated amphetamines and the ring-substituted methylenedioxy compounds. Three of the most popular routes are the Leuckart method (a non-metal reduction), reductive amination (a catalytic metal reduction), and synthesis *via* a nitrostyrene intermediate.

The Leuckart method is usually carried out in three steps (Figure 8.2). A mixture of the starting material, phenyl-2-propanone (P-2-P, phenylacetone) and formamide, sometimes in the presence of formic acid or ammonium formate, is heated by reflux, producing a reaction intermediate, N-formylamphetamine. N-formylamphetamine is hydrolysed using HCl, resulting in amphetamine. The product is then isolated using a series of acid/base extractions and/or distillation of the amphetamine-free base from the reaction mixture. This reaction results in a wide variety of reaction intermediates and by-products that contaminate the amphetamines, including N-formylamphetamine, a number of phenylpyrimidines, benzylpyrimidines, and phenylisopropylamines (Figure 8.3). These compounds occur at concentrations downward of 10^{-1} to 10^{-3}% of the amphetamine concentration in the sample and so special preparation is required if they are to be used in sample comparison (Section 8.9).

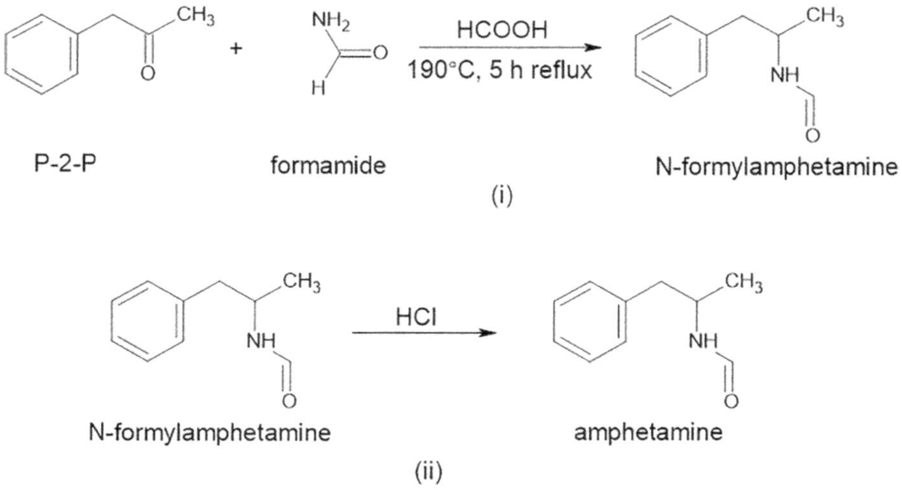

Figure 8.2 The first two steps in the Leuckart synthesis of amphetamine: (i) the reaction of P-2-P with formamide to form N-formylamphetamine and (ii) its subsequent hydrolysis in HCl. Amphetamine (usually as the hydrochloride salt) is then recovered from the reaction mixture using a series of acid/base extractions.

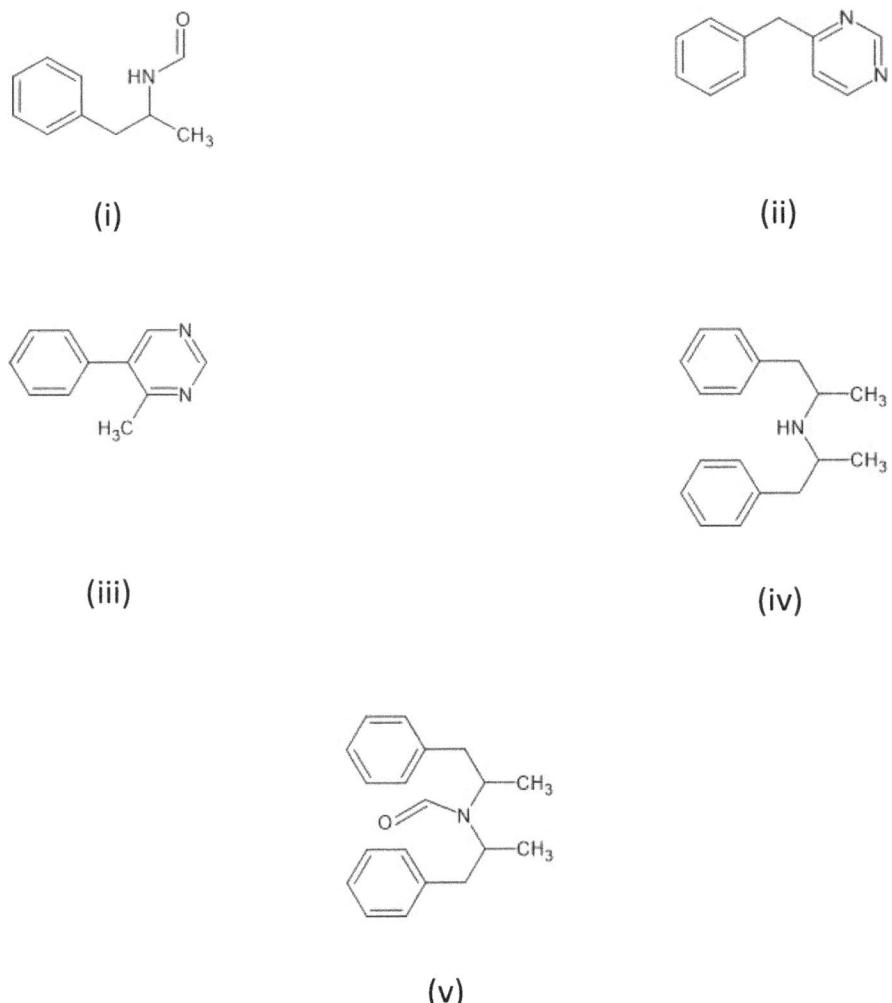

Figure 8.3 Examples of reaction intermediates and impurities formed during the Leuckart synthesis of amphetamine: (i) N-formylamphetamine, (ii) 4-benzylpyrimidine, (iii) 4-methyl-5-phenylpyrimidine, (iv) N,N-di(β-phenylisopropyl)amine, and (v) N,N-di(β-phenylisopropyl)formamide.

Reductive amination involves the catalytic reduction, in the presence of a metal catalyst, usually platinum or mercury, and under conditions of high temperature and pressure, of phenyl-2-propanone in the presence of ammonia (forming amphetamine), methylamine (forming methamphetamine), or further higher alkylamines. This

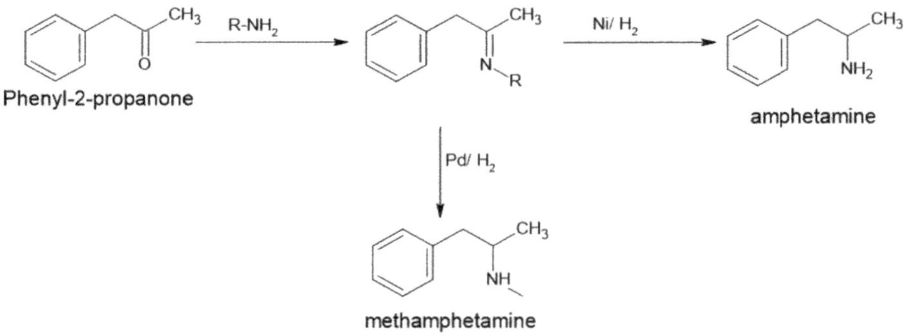

Figure 8.4 Reaction sequence for the formation of amphetamines by reductive amination.

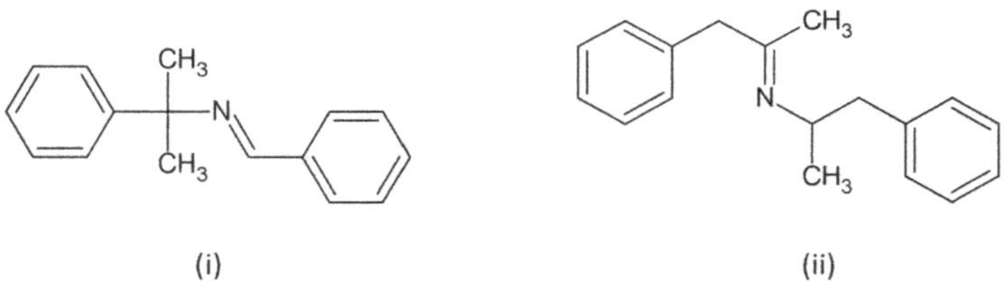

Figure 8.5 Structure of (i) *N*-(β-phenylisopropyl)-benzaldimine and (ii) *N*-(β-phenylisopropyl)benzylmethylketimine.

results in the formation of an imine intermediate product, which is subsequently reduced, forming amphetamine or an equivalent product (Figure 8.4). Reaction by-products include *N*-(β-phenylisopropyl)-benzaldimine and *N*-(β-phenylisopropyl)benzylmethylketimine (Figure 8.5).

The nitrostyrene route (Figure 8.6) involves the reaction of benzaldehyde and nitroethane in the presence of ammonium acetate and acetic acid under reflux conditions. This leads to the reaction intermediate, nitrostyrene, which can easily be crystallised from the reaction mixture. The nitrostyrene is subsequently reduced with lithium aluminium hydride or in a reaction that can be more easily controlled with sodium cyanoborohydride. Reaction by-products that form as a consequence of this route include nitrostyrene itself, methylphenylaziridines, and benzoylmethylketoxime.

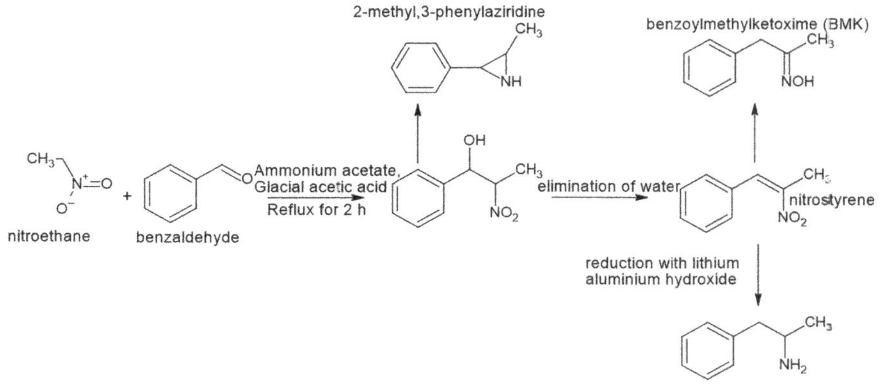

Figure 8.6 Synthesis of amphetamine by the nitrostyrene route and origins of aziridine and ketoxime impurities.

8.3 Description and Sampling of Amphetamine-type Drugs

Amphetamine drugs will be encountered by forensic scientists in a wide variety of forms. The way that these are treated will depend on what the form is. In each case, the materials should be divided into visually identical samples and the trace materials, for example, paraphernalia, should be separated. The number of visually identical materials should then be determined. Where the number is small, they can simply be counted. Where the number is large, for example, seized tablet dose forms, the weight rather than the number might be used to determine the number of dose units. To estimate the number, the total weight of the seizure should be divided by the average weight of a small number of tablet doses. The appropriate number of samples should then be taken, as described in Chapter 2.

Once the materials have been separated and sampled, the next step is to take materials for chemical analysis. For powdered samples, the material should be taken, as described in Chapter 2. Where the sample is a liquid, a small amount should be taken, ensuring that it is representative of the bulk liquid from which it is drawn. Where there are two or more phases to be sampled, a small amount of each phase should be taken for analysis.

Tabletted dose forms deserve special consideration. Often, there will be break marks or logos on the tablet(s), which may be a trademark of the drug manufacturer or perhaps designate the drug that is contained within the tablet. Additionally, there may be unique toolmark characteristics on some of the tablets resulting from damage to the die in which the tablet was pressed. These are referred to as the "ballistics" of the tablet, and their examination is not dissimilar to that used in firearms or toolmark examination. Where a number of tablets have been pressed in the same die, the damage will be identical. Such damage can be used for the comparison of samples within a batch and indeed between batches. Therefore, when taking powdered material from a tablet, these areas should be avoided in order to preserve the evidence for further examination or testing by another examiner. Where a tablet is being analysed, it should be remembered that the tablet should not be ground up in its entirety because this will destroy the physical evidence that the tablet represents.

Amphetamine-type drugs may also form trace evidence. For example, glassware used to smoke methamphetamine or items used to inject amphetamine may be presented to the forensic scientist. In such cases, the materials and the paraphernalia should be described and the drugs should be sampled according to the trace sampling protocol in Chapter 2.

8.4 Presumptive Tests for Amphetamine-type Drugs

Once the materials have been described and sampled, the next step in the analysis of amphetamines is to determine which of this group of drugs are present in the samples. The best screening method to do so is to use a presumptive test. Two tests that are commonly used during this process are the Marquis reagent and Simon's reagent.

In order to test with the Marquis reagent (100 ml of concentrated sulfuric acid mixed with 5 ml of 40% formaldehyde), a small drop of reagent is added to a test substrate, a known sample of the drug, and a negative control in a porcelain dimple well tray. The Marquis reagent undergoes a yellow to orange reaction with amphetamine and methamphetamine. The intensity of the colour reaction and the shade of orange depend on the amount of drug present in the sample. The more of the drug that is present in the sample, the more intense the orange colour reaction. However, this test should in no way be regarded as a quantitative test – in order to determine the amount of this type of drug present in a sample, an instrumental method is required. Other drugs, for example, fentanyl, will also give an orange colour reaction. The Marquis reagent can also be used to identify ring-substituted amphetamines, including MDA and MDMA. With ring-substituted amphetamines, the Marquis reagent gives a purple-blue colour. This should not be confused with the colour obtained when the Marquis reagent reacts with opiate drugs and illustrates the importance of using positive controls when using presumptive tests to determine which drug classes are present in a sample. While the Marquis reagent will differentiate between amphetamines and ring-substituted amphetamines, it will not differentiate between primary amines and *N*-methylated amphetamines. In order to achieve this distinction, the analyst should use Simon's reagent. Simon's reagent comprises two solutions. The first is a mixture of 2% (w/v) sodium nitroprusside and 2% (w/v) acetaldehyde in water. The second is a solution of 2% (w/v) sodium carbonate in water. The test is performed by placing one drop of the first reagent on the substrate followed by a drop of the second. This reagent gives a blue colour with secondary amines (the *N*-methylated drugs) but not with primary amines. As a consequence, using these two presumptive tests, it is possible to discriminate between the most common of the amphetamine-type drugs that a forensic scientist is likely to encounter.

Another test, the gallic acid test, allows the forensic scientist to discriminate between ring-substituted amphetamines, for example, MDMA, MDA, and MDEA, from amphetamine or methamphetamine. The gallic acid test reagent (50 mg of gallic acid dissolved in 10 ml of concentrated sulfuric acid) reacts with the 3,4-methylenedioxy substitution on the benzene ring. This means that with impurities, including starting reagents such as iso-safrole and safrole, there will also be a positive reaction in the gallic acid test. On this basis, these tests should be used in combination to determine the tentative identity of any materials suspected to contain amphetamine-type drugs.

8.5 Thin Layer Chromatography of Amphetamine-type Drugs

A number of thin layer chromatography systems are available for the analysis of amphetamines. In order to analyse amphetamine-type substances by thin layer chromatography, it is necessary first to dissolve them. A solvent that will dissolve all of the drugs of interest without causing decomposition or reacting with them should be chosen. The United Nations Office of Drug Control recommends methanol for this purpose. The methanol should be dry (anhydrous sodium sulfate can be used to dry the solvent), and the solutions should be stored in the dark at refrigerated temperatures. Once the street samples have been dissolved along with any standards, the solutions should be centrifuged to remove any particulate material. The presence of solid materials can lead to distorted chromatograms and incorrect drug identifications being made. Amphetamine-type drugs may, for example, contain starch, which has been used as a bulking agent that will not be soluble in methanol. Other materials that are insoluble include substances such as stearates used during the tableting process of amphetamine-type drugs. The stearates improve the flow properties of the powders containing the drugs into the tableting machine.

Having dissolved the drugs and removed any particular material, the next step is to prepare the chromatographic system. The stationary phase is typically silica gel containing a fluorescent indicator that fluoresces at 254 nm. A number of solvent systems have been used for the analysis of amphetamines. These include (by volume)

- methanol/concentrated ammonia 100/1.5
- ethyl acetate/methanol/concentrated ammonia 85/10/5
- cyclohexane/toluene/diethylamine 75/15/10 and
- methanol/acetone/concentrated ammonia 25/6/0.4.

In general, there is no need to convert the drugs to the free base form because the solvent systems contain concentrated ammonia or diethylamine, which converts the drugs to the free base form in the chromatographic process. This reduces the tailing of the amphetamines on silica gel, aiding the determination of the potential identity of the drug. When analysing amphetamine-type drugs by thin layer chromatography, positive and negative controls should also be prepared. The substances to be contained in the positive control will have been determined using presumptive tests. The negative control is methanol (or the solvent that has been used to prepare the sample) alone, which demonstrates that the drugs came from the sample and not from the solvent.

The thin layer chromatogram is prepared for the sample(s), positive control(s), and negative control spots. The spots are allowed to dry, and the chromatogram is developed in the mobile phase. Once the chromatogram has developed, it is removed from the chromatographic tank and air-dried. It should not be heated to hasten the drying process because the drugs are now in the freebase form and are far more volatile. Once the chromatogram has been dried, it should be viewed under (i) white light, (ii) short wavelength UV light (254 nm) where any compounds visualised will be observed as purple spots on a fluorescent background, (iii) long wavelength light where compounds will fluoresce and (iv) after spraying with a visualisation reagent. Typical visualisation reagents include 0.5% Fast Black K in water, which gives yellow, orange and red spots; ninhydrin (purple spots); or acidified potassium iodoplatinate (a variety of colours).

The R_f value and colour reaction of materials in the drug samples should be compared to those of the positive controls, and using such comparisons, tentative identification of the drugs can be made. It is important to compare the data from standards and samples on the same chromatographic plate because the R_f values obtained will depend on the degree of saturation of the TLC tank by the mobile phase, the temperature of the room, the age of the mobile phase and differences between batches of TLC plates.

8.6 High-performance Liquid Chromatography of Amphetamine-type Drugs

High-performance liquid chromatography is a useful technique for the analysis of amphetamines, which are highly polar and generally require derivatisation before GC or GC–MS analysis. Amphetamines can be both identified and quantified by HPLC. In addition, some jurisdictions require the identification of the isomers in the sample, and using either crown ether stationary phases or derivatisation, it is possible to differentiate between stereoisomers.

The simplest systems are normal-phase and reversed-phase systems. A simple normal-phase system involves dissolving the amphetamines in methanolic HCl (100 ml of methanol in which 175 µl of HCl has been dissolved). This causes the formation of the hydrochloride salt of the drug, which supports the ion-exchange separation mechanism on which the separation mechanism is based. Compounds including amphetamine, methamphetamine, MDA and MDMA can be separated using a silica gel column (12.5 cm × 4.6 mm i.d., 5 µm particle size), a mobile phase of MeOH/NH_3/HCl (2000/18.4/5.8 by volume), a flow rate of 1 ml min^{-1}, with detection at

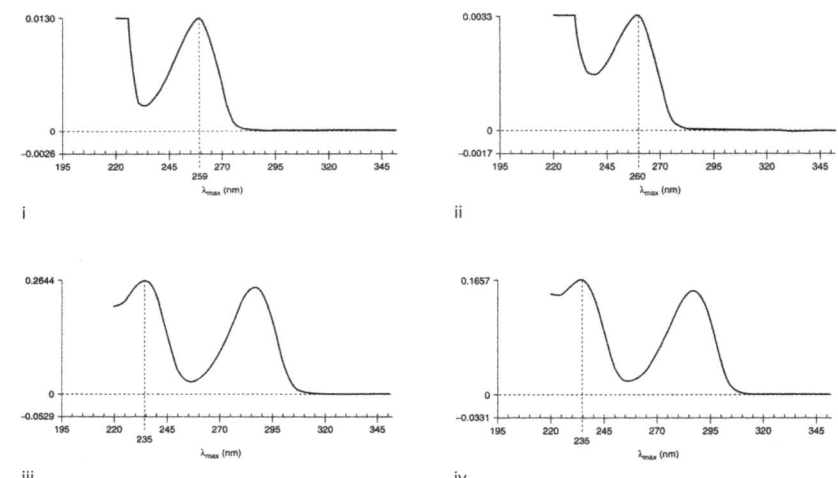

Figure 8.7 Diode array spectra of (i) amphetamine, (ii) methamphetamine, (iii) 3,4-methylenedioxyamphetamine (MDA) and (iv) 3,4-methylenedioxymethamphetamine (MDMA) in the HPLC system described. Amphetamine and MDA, and methamphetamine and MDMA have the same retention time because of the separation mechanism but can be differentiated on the basis of their diode array spectra.

254 nm or with diode array detection. Since the separation process in this system is based on ion exchange with the primary and secondary amines, both the amphetamine/MDA (primary amine) and methamphetamine/MDMA (secondary amine) have the same retention time. However, they can be discriminated on the basis of their diode array spectra (Figure 8.7).

Reversed-phase systems have also been described. One such example[2] has been used to separate pseudoephedrine, ephedrine, MA, MDA, MDMA and MDEA. Samples were dissolved in a phosphate buffer at pH 7, centrifuged and analysed directly on an ODS column (25 cm × 1.5 mm i.d.) in a mobile phase of 5 mM SDS in 20 mM KH_2PO_4/acetonitrile (65/35) at a flow rate of 0.1 ml min^{-1}. The column temperature was 50 °C with UV detection at 210 nm. The elution order could be predicted on the basis of compound polarity and the separation mechanism of partition, with pseudoephedrine and ephedrine eluting before methamphetamine, and the ring-substituted 3,4-methylenedioxy compounds eluting in the order MDA, MDMA, and MDEA.

It is sometimes necessary to achieve separation of the isomers of amphetamines. Different isomers of amphetamine have different activities – the 2S isomer of amphetamine exhibits 10–20 times

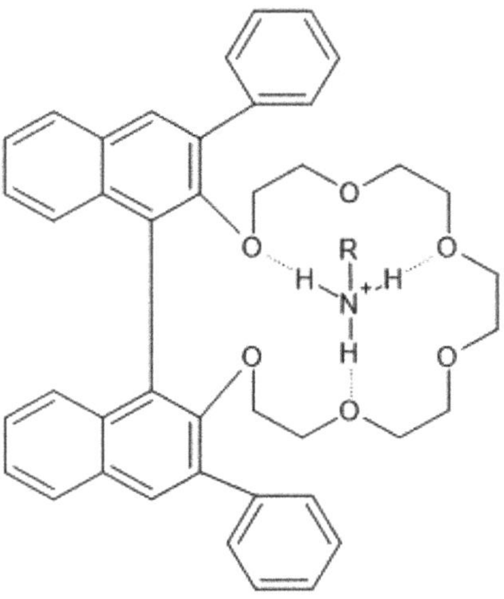

Figure 8.8 Schematic of the intercalation of an ammonium ion with the crown ether of a Daicel Crownpak CR(+) column. Adapted from ref. 3 with permission from Elsevier, Copyright 1996.

more stimulant activity than the 2R isomer. This may be required to determine whether or not the samples were licitly (usually racemically pure) or illegally synthesised (usually a racemic mixture) or to determine the relative proportions of isomers to compare samples. One mechanism by which separation by HPLC may be achieved is using crown ethers in the stationary phase. One system has used a 15 cm × 0.4 mm i.d. Daicel Crownpak CR(+) column[3] (Figure 8.8) with a mobile phase of $HClO_4$ (pH 1.8) at a flow rate of 1 ml min^{-1}, with UV detection at 200 and 254 nm. Using this system, it is possible to separate isomers of norephedrine (1S, 2R and 1R, 2S norephedrine) and amphetamine (R-amphetamine and S-amphetamine). The system has the advantage that derivatisation is not necessary, excluding the possibility of incomplete derivatisation or artefact formation. However, the system does not separate the isomers of methamphetamine. At the pH used (pH 1.8), the amphetamine, for example, will be in the cationic form. The cations interact to a different extent with the crown ether depending on the stereochemistry of the drug (Figure 8.8). In the case of both norephedrine and amphetamine, the 2S compound (1R, 2S norephedrine and 2S amphetamine) eluted before the 2R equivalent because the 2R compounds interact more strongly with the crown ether. The two isomers of methamphetamine do not separate because the N-methyl

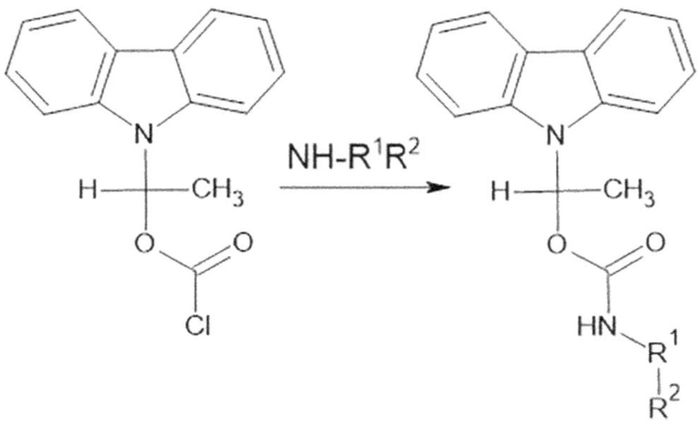

Figure 8.9 Reaction of primary and secondary amines with (−)-1-(9-fluorenyl)ethyl chloroformate prior to HPLC analysis.

group prevents complex formation with the crown ether, and hence, they elute together.

When simple mixtures occur, it is also possible to separate isomers of primary and secondary amphetamines with, for example, (−)-1-(9-fluorenyl)ethyl chloroformate (FLEC).[4] It is widely known that chloroformates react with both primary and secondary amines, which is an important consideration when analysing amphetamine-type mixtures (Figure 8.9). The derivatives are stable and fluoresce, the latter offering sensitivity for the method. However, there are some problems with the method. Sometimes, baseline resolution of the isomers could not be achieved (for example, amphetamine and methamphetamine), which makes quantification difficult. Additionally, retention times of some groups of compounds, *e.g.* MDMA and MDEA, were similar, making definitive identification difficult. In addition, the reagent needs to be in molar excess to ensure that the drugs are fully derivatised. Conditions on autosamplers need to be identified so that the derivatives are stable. On this basis, the use of derivatisation offers the opportunity to separate some isomeric pairs of some amphetamines but not all. The quality of the derivatisation, chromatography and drug identification needs to be determined experimentally.

8.7 Gas Chromatography of Amphetamines

There are a wide range of techniques available for the gas chromatographic analysis of amphetamines. These include qualitative methods where underivatised amphetamines are analysed and quantitative methods where amphetamines are determined, usually following derivatisation, by gas chromatography–mass spectrometry.

8.7.1 Qualitative Analysis of Underivatised Amphetamines

The United Nations Office for Drug Control recommended methods for amphetamines[5] include a qualitative technique where underivatised amphetamines are analysed by gas chromatography with FID or NPD detection. Street samples and standard amphetamine-type substances should be dissolved in water at a concentration of 1 mg ml^{-1}. To the solution, a 5% solution of sodium hydroxide is added dropwise to increase the pH to 10 and convert the drugs to their free base form. These drugs can then be analysed following extraction into an organic solvent, which may be hexane, chloroform or methylene chloride. The solvent should be sufficiently polar to dissolve the drugs, in their free base forms, quantitatively. It should also be chosen so that the drugs are stable and do not form artefacts with it. It is for this reason that solvents containing ketones should be avoided. The solvent should also be compatible with gas chromatographic analysis. The gas chromatographic conditions that can be used are given in the recommended methods. Identification of the compounds is achieved by comparison of the retention time of the drug with the retention time of a standard. Caution should be exercised in the interpretation of data if this method is used. FID and NPD will not provide definitive identification of the drugs. While retention time may be used to tentatively identify a drug, there may be problems of co-elution that would not be detected using this method. As a consequence, this method, while providing a tentative identification, is not definitive because there are no spectroscopic data to support the compound identification. Second, the method cannot be used as a quantitative method because there is no internal standard. Finally, the use of butyl acetate and methylene chloride or chloroform must be very carefully considered. Primary amines react with ketones, and thus, artefacts may be formed with butyl acetate. Methylene chloride and chloroform may contain hydrochloric acid, which may cause some compounds to decompose. Finally, an inconvenient emulsion may form between any aqueous layer and the chlorinated hydrocarbons during the extraction process. For these reasons, the methods described below may be more appropriate for the definitive identification of amphetamines.

8.7.2 Identification of Amphetamines by GC-MS

The most widely used method for the definitive identification of amphetamines is GC-MS. Using this technique, amphetamines are ideally derivatised prior to analysis because the mass spectra of many of the amphetamines under electron impact conditions are very difficult to discriminate with very similar fragmentation patterns (ions at m/z = 119, 91, 65 and 51) and relatively small amounts of parent ions or sometimes none are present at all. Additionally, derivatisation improves the chromatographic behaviour of the amphetamines, which are very polar and subject to tailing and sorption onto the chromatographic system.

Detection of amphetamines following derivatisation is much improved. This is particularly important in some samples where the concentration of the drug may be less than 5%, which means that the concentration of the freebase of, for example, amphetamine is less than 2.8%.

One method that is widely used in the United Kingdom is the derivatisation of primary amines using carbon disulfide (Figure 8.10) to make the NCS derivative. For bulk samples, this is achieved by dissolving approximately 2 mg of the drug in a small amount of ammoniacal methanol. The ammonia is added so that the free base of the drug is released into the solution – this is necessary for the reaction with carbon disulfide. A small amount of carbon disulfide is added to the mixture, which is thoroughly shaken and allowed to stand for at least 30 min at room temperature. The solvent is evaporated under nitrogen. Nitrogen is used because it is dry and inert, and the mixture is re-suspended in methanol and analysed by GC-MS. Identification is made on the basis of retention time and the mass spectrum of the NCS derivative of amphetamine (Figure 8.11). Additionally, a positive (known amphetamine sample) and a negative control (solvent containing no drug) should also be derivatised and analysed in order to demonstrate that any amphetamine type of substance identified has come from the street sample and from nowhere else. The formation of the NCS derivative renders the amphetamine less polar, improving its chromatographic behaviour and, at the same time, increasing the sensitivity of the method.

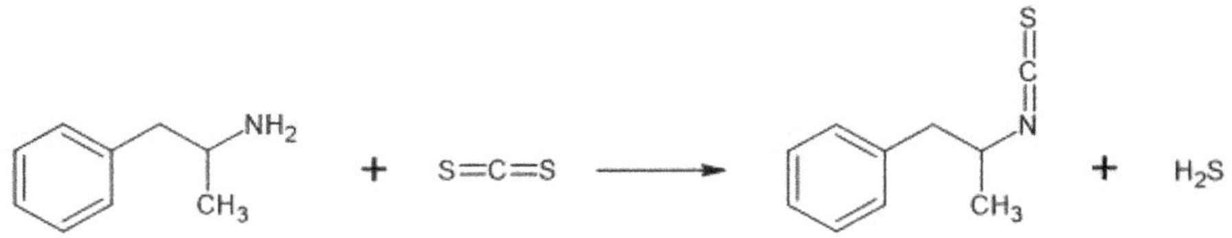

Analysis of Amphetamines and Related Compounds

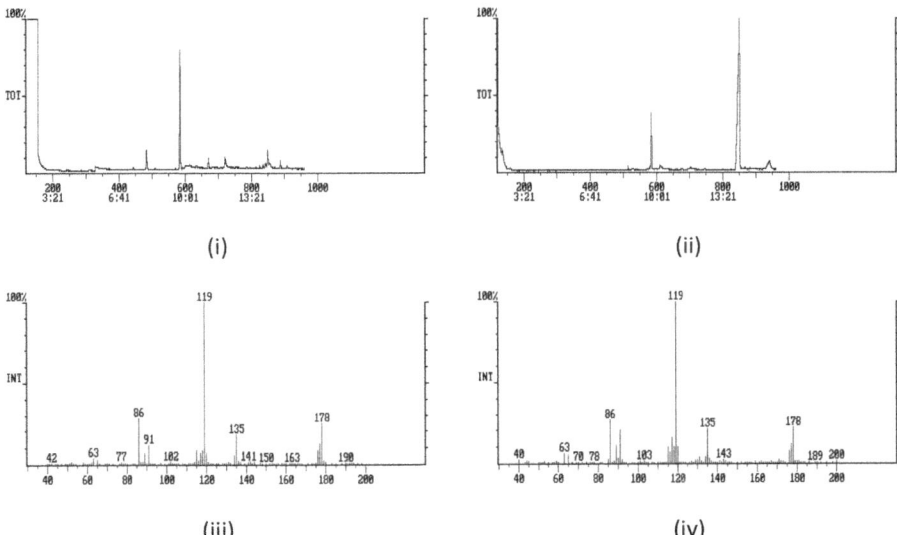

Figure 8.10 Reaction of amphetamine with carbon disulfide to form the "NCS" derivative prior to GC–MS analysis.

Figure 8.11 Identification of amphetamine in a street sample by GC–MS after NCS derivatisation. (i) Gas chromatographic separation of standard amphetamine derivatised with CS_2 (R_t = 9 min 45 sec) (note that reaction artefacts are also formed), (ii) gas chromatographic separation of the street sample of amphetamine derivatised with CS_2 (amphetamine R_t = 9 min 45 sec, caffeine = 14 min 5 sec), (iii) mass spectrum of the NCS derivative of standard amphetamine and (iv) mass spectrum of the NCS derivative of amphetamine in the street sample.

A problem with this method of derivatisation is that while it works for primary amines, it does not work for secondary or tertiary compounds and another derivatisation method is required. A derivatisation technique that can be used for the majority of amphetamine compounds, including primary and secondary amines, involves the use of acid anhydrides (Figure 8.12). One of the most commonly used acid anhydrides is heptafluorobutyric anhydride (HFBA). Samples are typically converted into their free base

(i) (ii)

Figure 8.12 Structures of the HFBA derivatives of (i) amphetamine and (ii) methamphetamine.

form using an aqueous buffer at a high pH. The non-polar free bases are then extracted into ethyl acetate, and a small amount of HFBA is added. The derivatisation mixture is gently heated (65–70 °C) for 30 min. After derivatisation and cooling, the samples are evaporated to dryness and reconstituted in ethyl acetate containing an internal standard. Samples are identified on the basis of the retention time or retention index of the derivatised sample and the comparison of the mass spectrum of the derivatised sample against the derivative of a positive control.

8.8 The Analysis of Inorganic Ions in Samples of Amphetamine and Related Drugs

Metal ion analysis can be used in a number of ways in amphetamine analysis. The body of literature is not large, but there are some interesting examples where such analyses have been applied to MDMA samples. Metal ions arise from a number of sources, including reducing agents used during drug synthesis (Al, B, Hg, Li), catalysts (Ni, Pd, Pt), colouring agents (Cu, Co, Fe), adulterants and diluents (Ba, Ca, Cd, Fe, Mg, Na, Pb), and from leaching from metal reaction vessels.[6] Of the techniques available for metal ion analysis, ICP-MS offers the greatest sensitivity to the widest range of metal ions although other techniques for metal determination can be applied. Samples are typically prepared using digestion in nitric acid and hydrogen peroxide, either at atmospheric pressure at temperatures of 40 °C and 90 °C or using microwave digestion at temperatures around 280 °C at 70 bar (1015 PSI). After cooling, the samples are typically diluted with water or an internal standard solution. The samples can then be analysed by ICP-MS.

In drawing conclusions about the metal ion content, it is necessary to understand the variation in concentrations likely to be expected in samples. Work on intra-batch variation in metal ion concentration has shown that the variation in powder samples is greater (8–25%) than in tablet dose forms.[6] The mixing and grinding in addition to the smaller particle size of the powders prior to tableting, compared to powder samples that are then used to ingest the drug in other ways, lead to greater homogeneity within the sample.

It is also possible to identify the synthetic route through inorganic analyses. For example, the synthesis of MDMA in some parts of the world is by one of three methods. The first is the "high-pressure" method, with Pt as the catalyst. Samples synthesised by this route typically have Pt concentrations of 6–2182 ppm,[6] with B and Hg typically absent. The "cold method" uses $NaBH_4$ as the reducing agent and contains B at concentrations of 6–48 120 ppm, with Pt and Hg typically absent. Finally, the "aluminium amalgam reduction method" uses $HgCl_2$ and results in mercury concentrations of 0.2–290 ppm, with Pt and B typically not detected.[6] The presence of one of the three ions in such samples can be used to identify the synthetic route employed. Where inorganics occur together, it may be that the samples have come from more than one source batch. For example, in samples where B and Pt were found to occur together, the authors opined that the samples had been mixed, with one sample synthesised by the high-pressure method the other by the cold method. Since speciation is not required for such analyses, inorganic ion analysis may offer advantages over the analysis of the organic impurities and reaction intermediates in amphetamine samples for the identification of the synthetic routes and comparison of samples. This is because organic molecules are subject to oxidation and hydrolysis reactions, which will change the proportions of these molecules in the sample. Oxidation or hydration of inorganic ions will not result in the same significant change.

8.9 Chemical Impurity Profiling of Amphetamine-type Drugs

There is considerable history and body of literature relating to the chemical impurity profiling of amphetamine-type drugs and most use the organic reaction impurities and intermediates for the profiling process. There are, however, very few profiling methods that have been systematically developed for profiling these groups of drugs. This is, in part, due to (i) the time required to develop such methods and (ii) the cost involved. One example where there has been a systematic study is that of amphetamine itself.[7-12] In developing such methods, a number of principles should be adhered to so that the profiling method is accurate, precise, repeatable and robust.

At the outset, it is important that any method for chemical impurity profiling is sufficiently sensitive to detect all of the significant impurities. This is because some of the impurities occur at very low concentrations (down to 10^{-3}% or less) relative to the drug in which they are found, with the drug itself occurring at low concentrations (1–5%) in the street sample. Increased sensitivity can sometimes be achieved by derivatisation prior to analysis, but this has the associated problem around assuming that all of the sample has been derivatised (and hence a molar excess is required) and that the reaction itself is both clean (and does not produce further impurities which would "crowd", for example, any chromatographic output), and does not result in artefact formation. It is for this reason that the derivatisation for chemical impurity profiling is best avoided.

It is also necessary to understand which impurities might be present in a chemical profiling sample. This can be explored using "paper chemistry" and anticipating the impurities that might be present. For a complete systematic study, synthesis, identification, generation of analytical data and documentation of the potential impurities are then required, as exemplified for amphetamine.[8] This may require the synthesis of the impurities because they may not be commercially available. The identity of each should then be confirmed, and analytical data should be collected. Examples of the types of data needed are given in Table 8.2, using N,N-di-(β-phenylisopropyl)amine as the exemplar impurity.

Having synthesised and characterised potential impurities, it is necessary to determine the best solvent for the extraction of the impurities from the drug sample. The impurities may have a wide range of polarities, and so extremely non-polar (hexane) and very polar (water) are generally not suitable. Additionally, the solvents should not react with the potential impurities, which precludes water (hydrolysis) and ketones (reaction with amine groups), for example. Chlorinated solvents may not be suitable for some impurities due to UV-catalysed breakdown, leading to HCl formation in the solvent. Some impurities will be acid-labile, and hence, the HCl will catalyse their breakdown. The solvent choice should be made carefully, using both paper chemistry and experimentation.

Experimentation may be required because unexpected breakdown of compounds may occur – for example, the decomposition of some impurities found in amphetamines in isooctane.[9] Storage conditions should also be investigated. It is known that, for example, some impurities from the Leuckart synthesis are unstable in daylight because they break down in UV light.[13] It is for this reason that they should be stored in the dark and/or amber-coloured containers. Where compounds are found to be unstable and this cannot be rectified, it has been recommended that these compounds are not used in chemical impurity profiling or sample comparison.

Once the putative impurities have been synthesised and a reliable system is established for their storage and dissolution, instrumental parameters can then be optimised. The optimisation should take account of all of the potential routes of synthesis and the many different classes of molecules that result. Additionally, each part of the instrumental method should be optimised, including how the sample is introduced to the instrument, how the components of the mixture are separated and the detector type and the settings that are used for the detector. Such experiments are exemplified by those that have been undertaken on the impurities found in amphetamines.[10]

Once the analytical method has been optimised using standard impurities, it is possible to return to the optimisation of the extraction method. This is because it is then known that impurities are stable in the chosen solvent and that the responses of the analytical instruments are accurate and precise. The method should result in the quantitative and complete extraction of the impurities from the

Table 8.2 Exemplar reference data that might be generated following the synthesis of a suspected impurity in an amphetamine sample, using N,N-di-(β-phenylisopropyl)amine as the example. The full dataset can be obtained from ref. 8.

Mass spectrometry	m/z = 162 (100%), 91, 119, 44
IR (cm^{-1})	2962, 744, 2924, 1451, 3025, 1493, 703, 1373, 1141, 1342
^{1}H NMR (400 MHz)	δ = 0.95–1.05 (3H, d, J = 6.1 Hz, CH$_3$), 2.51–2.68 (2H, dd, J = 6.9 Hz, CH$_2$), 2.83 (1H, dd, J = 6.1 Hz, CH), 3.04 (1H, q, CH), 7.2 (10H, m, aromatic).
UV (λ max) (nm) (MeOH)	204, 252, 260

drug sample for analysis so that the profile is truly representative of the drug sample being analysed. Again, both liquid-liquid extraction (LLE) and solid phase extraction (SPE) methods should be considered as exemplified for amphetamines.[11]

Finally, the variability of the method needs to be understood, as exemplified for amphetamine.[12] Where numerical classification schemes are used, the underlying assumptions around the datasets (for example, normality of the data, independence or correlation of the data *etc.*) should be appropriately tested. Only when the underlying assumptions are met, should the tests be used.

There are difficulties when using the approach described above to develop a method for chemical impurity profiling. If a new synthetic route appears, with new chemical families of impurities, the process needs to be repeated, especially if data are to be exchanged between laboratories. Some jurisdictions do not have the resources to undertake such detailed studies. Finally, new classes of drugs, including amphetamines, continue to appear. The consequence of this is that the appropriateness of any profiling method needs to be subject to constant review.

References

1. A. Shulgin and A. Shulgin, in *PIHKAL: A chemical love story*, Transform Press, Berkeley, California, 1991.
2. Y. Makino, S. Tanaka, S. Kurobane, M. Nakauchi, T. Terasaki and S. Ohta, *J. Health Sci.*, 2003, **49**(2), 129.
3. Y. Makino, S. Ohta and M. Hirobe, *Forensic Sci. Int.*, 1996, **78**(1), 65.
4. P. J. Campins Falco, J. Verdu-Andres and R. Herraez-Hernandez, *Chromatographia*, 2003, **57**, 309.
5. Available from: www.unodc.org/unodc/en/scientists/recommended-methods-for-the-identification-and-analysis-of-amphetamine--methamphetamine-and-their-ring-substituted-analogues-in-seized-materials.html. Accessed 19 Aug 2024.
6. C. Koper, C. Van Den Boom, W. Wiarda, M. Schrader, P. De Joode, G. Van Der Peijl and A. Bolck, *Forensic Sci. Int.*, 2007, **171**(2-3), 171.
7. J. Ballany, B. Caddy, M. Cole, Y. Finnon, L. Aalberg, K. Janhunen, E. Sippola, K. Andersson, C. Bertler, J. Dahlen and I. Kopp, *Sci. Justice*, 2001, **41**(3), 193.
8. L. Aalberg, K. Andersson, C. Bertler, H. Borén, M. D. Cole, J. Dahlén, Y. Finnon, H. Huizer, K. Jalava, E. Kaa and E. Lock, *Forensic Sci. Int.*, 2005, **149**(2-3), 219.
9. L. Aalberg, K. Andersson, C. Bertler, M. D. Cole, Y. Finnon, H. Huizer, K. Jalava, E. Kaa, E. Lock, A. Lopes and A. Poortman-van der Meer, *Forensic Sci. Int.*, 2005, **149**(2-3), 231.
10. K. Andersson, K. Jalava, E. Lock, Y. Finnon, H. Huizer, E. Kaa, A. Lopes, A. Poortman-Van der Meer, M. D. Cole, J. Dahlén and E. Sippola, *Forensic Sci. Int.*, 2007, **169**(1), 50.

11. K. Andersson, K. Jalava, E. Lock, H. Huizer, E. Kaa, A. Lopes, A. Poortman-Van der Meer, M. D. Cole, J. Dahlén and E. Sippola, *Forensic Sci. Int.*, 2007, **169**(1), 64.
12. E. Lock, L. Aalberg, K. Andersson, J. Dahlén, M. D. Cole, Y. Finnon, H. Huizer, K. Jalava, E. Kaa, A. Lopes and A. Poortman-van der Meer, *Forensic Sci. Int.*, 2007, **169**(1), 77.
13. C. S. L. Jonson and N. Artizzu, Factors influencing the extraction of impurities from Leuckart amphetamine, *Forensic Sci. Int.*, 1998, **93**(2–3), 99–116.

9 The Analysis of Barbiturate Drugs

9.1 Introduction

Barbiturates are synthetic drugs based on 5,5-substitutions of barbituric acid (Figure 9.1), which act as central nervous system depressants. The substitutions include straight chain (alkyl) and branched alkanes and alkenyl and aryl substitutions at the $5a$ and $5b$ positions. The actions of barbiturates can be described as short (*e.g.* pentobarbitone and cyclobarbitone), which act for up to approximately 2 h, intermediate (*e.g.* amobarbital), which act for between 3 and 6 h, or long-lasting (*e.g.* phenobarbitone), which last over 6 h. It is the short- or intermediate-acting barbiturates that are generally used as recreational drugs. Administered by injection as solutions, or orally as tablets, these drugs have been used medically as anxiolytics and sedatives and to prevent convulsions and seizures. In addition, they have been used in anaesthesia and for the treatment of epilepsy.

Barbiturates are generally encountered in the forensic science context in pill form or mixed with other drugs. These include heroin, where barbiturates act as anxiolytics, and methamphetamine, ring-substituted amphetamines and cocaine, where barbiturates counteracts the stimulant effect of the accompanying drug. Prescribed for anxiety, depression and insomnia they have, in general, been superseded in the healthcare context by benzodiazepines. As a consequence, they are less-frequently encountered in forensic science samples than they once were. They are also dangerous when used as recreational drugs because their therapeutic index is low – there is only a small difference between a therapeutic dose

---

Controlled Drug Analysis
By Michael D. Cole, Lata Gautam and Agatha Grela
© Michael D. Cole, Lata Gautam and Agatha Grela 2025
Published by the Royal Society of Chemistry, www.rsc.org

The Analysis of Barbiturate Drugs

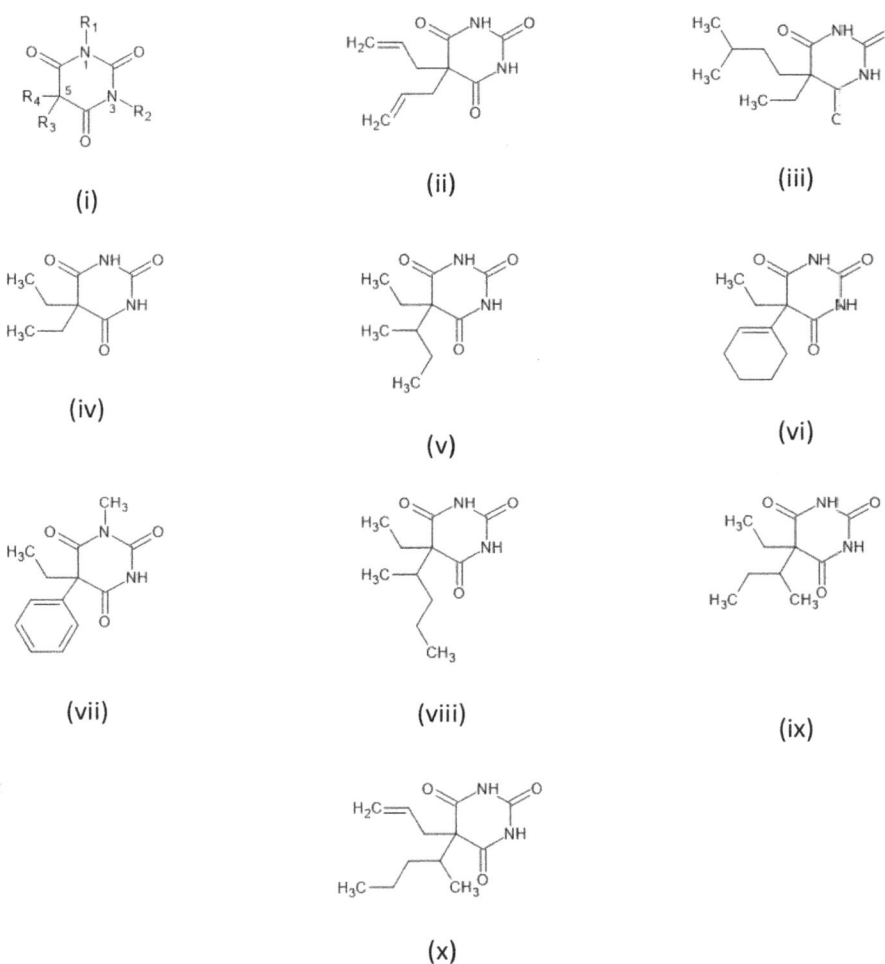

Figure 9.1 Structure of barbituric acid and exemplar 5,5 disubstituted barbiturates that are commonly encountered in forensic science. (i) Barbituric acid, (ii) allobarbital, (iii) amobarbital, (iv) barbital, (v) butabarbital, (vi) cyclobarbital, (vii) phenobarbital, (viii) pentobarbital, (ix) secbutabarbital and (x) secobarbital.

and a lethal one. Originally controlled in the UK by the Misuse of Drugs Act (Regulations) 1985, tabletted dose forms are controlled under the Medicines Act and are Class B drugs under the Misuse of Drugs Act 1971, where they are generically controlled as "any 5,5-disubstituted barbituric acid". Injectable forms carry more severe penalties as Class A drugs. As a consequence of some of

Table 9.1 Legislative classification of exemplar barbiturates illustrating that in some systems, the scheduling depends on medical use.

Drug	United Kingdom	United States	UN Convention on Psychotropic Substances (1971)
Allobarbital	B	III	IV
Amobarbital	B	II/III	III
Barbital	B	IV	IV
Butobarbital	B	III	IV
Cyclobarbital	B	III	III
Methylphenobarbital	B	IV	IV
Pentobarbital	B	II	III
Phenobarbital	B	IV	IV
Secobutobarbital	B	III	IV
Secobarbital	B	II/III	II

the barbiturates being used in medical practice, they are controlled differently according to their use (Table 9.1).

As powdered and tablet drugs, barbiturates are straightforward to sample and homogenise and standard protocols can be used to obtain a sample for analysis, as described in Chapter 2. For bulk samples, the standard sequence of analyses of presumptive tests, thin layer chromatography and instrumental analysis can be applied. For trace samples, the analyst should proceed straight to instrumental methods.

9.2 Presumptive Tests for Barbiturate Drugs

There are two tests available for screening of drug samples for barbiturates. For the first, the Dille–Koppanyi test, a mixture of two reagents is used. The first reagent is 0.1% cobalt(II) acetate in 100 ml of methanol mixed with 0.2 ml of glacial acetic acid and the second reagent is 5% (by volume) isopropylamine in methanol. Two drops of the first reagent and one drop of the second are added to the test substrate, and a purple colour is produced if barbiturates are present. The second test is the one using Zwikker reagent. Here, two solutions are used. The first is an aqueous solution of 0.5% copper sulfate (w/v) and the second is 5% pyridine in chloroform. One drop of each is added to the test substrate in order and a colour change is observed. Barbiturates react to produce a purple colour. As with all presumptive tests, positive and negative controls should be used to provide a reference colour change and demonstrate that the equipment and chemicals used for the test are free of contamination.

9.3 Thin Layer Chromatography of Barbiturate Drugs

TLC of barbiturates is typically carried out on silica gel, which contains a fluorescent indicator that fluoresces under UV light at 254 nm. Solvent systems include ethyl acetate/methanol/25% ammonia (85 : 10 : 5 by volume) or chloroform/acetone (4 : 1 by volume). Samples can be dissolved in methanol and any insoluble matter should be removed by centrifugation or filtration, and enough solution with approximately 10 µg of the sample or standard was applied to the chromatographic plate. Once developed, the chromatographic plate should be observed under UV light (254 nm), where the drugs will appear as purple spots on a fluorescent background. A specific spray reagent has been developed for barbiturates – mercuric chloride–diphenyl carbazone reagent. Two solutions are prepared: 0.2% diphenylcarbazone in ethanol and 2% mercuric chloride in ethanol. The solutions should be freshly prepared and mixed in equal volumes to form the reagent just before use. Barbiturates give a blue/violet colour reaction against a pink background, with a detection limit of 1–5 µg on a plate. Much care should be taken when using this reagent because of the toxicity of the mercury salts used. At the time of writing, a safer alternative has not been reported.

9.4 High-performance Liquid Chromatography of Barbiturates

While there are many HPLC systems available for the analysis of recreational drugs, the UNODC recommends simple reversed-phase chromatography as the method of choice for the analysis of barbiturates using a C-18 octadecyl silica column. Standard solutions of barbiturates should be prepared at a concentration of 1 mg ml^{-1} in methanol. Studies have shown that the standards are stable in methanol for up to 3 months in a freezer at -20 °C.[1] Sample solutions should be prepared in the same batch of methanol to give a barbiturate concentration of approximately 1 mg ml^{-1} and any insoluble material should be removed by filtration or, preferably, centrifugation. A methanol blank from the same batch of methanol used to dissolve the standards and samples should also be prepared to demonstrate that the methanol is free of contamination.

Barbiturates with short alkyl chains can be separated in a system using a 25 cm x 4.6 mm i.d. column of octadecyl silica (Spherisorb ODS-2 or equivalent) with a 5 µm particle size and a mobile phase of acetonitrile/water (30/70 by volume). An injection volume of 1–5 µl, a flow rate of 1 ml min^{-1} and UV detection at 220 nm can be used to separate and detect the analytes, which separate on the basis of the length and lipophilicity of the alkyl chain substitutions at C5. As the lipophilicity of the side chains increases, the proportion of acetonitrile can be modified and increased, facilitating the separation of the relevant barbiturates. Detection is at 220 nm because the molecules do not contain conjugated systems with, generally, only the double bond of the carbonyl groups available to facilitate detection.

The detection used for HPLC is not specific. Greater specificity can be achieved using absorbance ratios. By measuring the absorbance of the analytes as they elute at 220, 240 and 254 nm and calculating the ratios A_{220}/A_{254} and A_{240}/A_{254}, it is possible to increase the specificity of HPLC analysis of barbiturates.[2] However, it is still not possible to discriminate all of the drugs in this class and other methods are required.

9.5 Gas Chromatographic Analysis of Barbiturates

Gas chromatography of barbiturates can be carried out without derivatisation of the sample although the system needs to be kept very clean to avoid either the barbiturates sorbing onto the chromatographic system, resulting in reduced sensitivity, and/or poor chromatographic behaviour. One suitable system, a simple GC–FID method, has been described.[1] A capillary column GC is usually employed with a 25 m long BP-1 column (or equivalent) with an internal diameter of 0.32 mm and a layer thickness of 0.25–0.5 µm. The injection volume is usually 1 µl, the injection temperature is 275 °C, and split or splitless injections can be made depending upon the concentration of the barbiturates in the sample, with nitrogen as the carrier gas at a flow rate of 1 ml min^{-1}. The column oven can either be run isothermally (200 °C) or to improve the range of barbiturates that can be separated, and to improve chromatography, a column oven temperature programme can be used (200–260 °C rising at 4 °C min^{-1}). The drugs are identified on the basis of retention time alone. The problem with this system is that it is very generic, with an FID detector, which does not prove the identity of the drug, and as a consequence, the method cannot be regarded as definitive.

GC–MS systems have been developed using the equivalent of the GC system described above, hyphenated to a mass spectrometer. A system that can be used includes GC setup as mentioned above, with a transfer line temperature of 280 °C, an ion source temperature of 230 °C, an ionisation voltage of 70 eV operated in total ion chromatogram (TIC) mode with a scan range of 50–500 AMU.

The challenge posed by both the GC and the GC–MS methods described is due to the chemical nature of barbiturates and the problems associated with "sticking" to the GC system. One way that this may be overcome is by chemical derivatisation. Barbiturates lend themselves to a technique called flash alkylation. This is an "on-column" derivatisation technique where the derivatisation of the barbiturate does not occur until the injection of the sample solution into the injection port where the chemical reaction takes place (Figure 9.2). One reagent that can be used for this reaction is trimethylanilinium hydroxide dissolved in methanol. This reagent N-methylates the drugs, rendering them less polar and improving their chromatographic behaviour.

The sample is prepared in methanol as described above and mixed with the derivatisation reagent at a known concentration. If a quantitative analysis is required, an internal standard should also be included in the system. The mixture is then subjected to analysis. If GC–FID is employed, then the barbiturates are identified on the basis of retention time alone. If GC–MS is used, with the system described above, then any barbiturates in the sample are

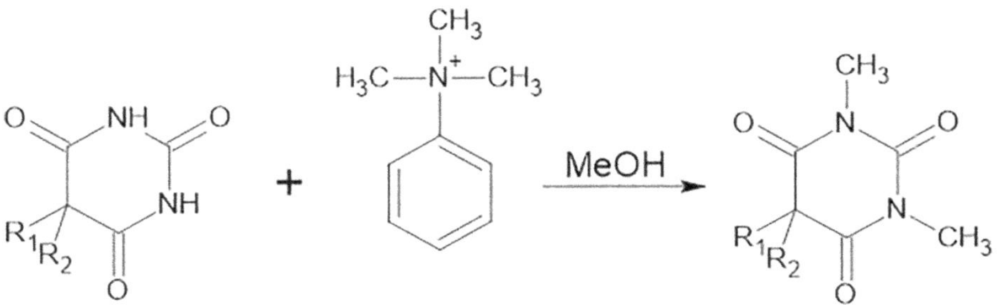

Figure 9.2 Reaction of a generalised barbiturate with trimethylaniliium [in the hydroxide form (TMAOH), in methanol] to form the N-methyl derivative.

identified on the basis of the retention time and mass spectrum of the N-methylated derivative of the native drug.

9.6 NMR Spectroscopy of Barbiturates

One method that can be used to overcome the chromatographic difficulties associated with the analysis of barbiturates is NMR spectroscopy. The technique allows the identification of both pure compounds and barbiturates in complex mixtures,[3] even in relatively low strength fields (100 MHz), with the sample dissolved in $CDCl_3$. Using this technique, the barbiturates can be differentiated on the basis of the signals associated with the substitutions at the C5 position, using both the downfield shift and integration values. For example, the NMR spectrum of phenobarbitone gives rise to a 3-carbon triplet associated with the methyl group, a 2-proton quartet associated with the methylene group, a 3-proton singlet associated with the N–CH$_3$ group, a 5-carbon multiplet associated with the aryl group and a broad 1-proton singlet associated with the N–H group. Occasionally, the crystalline nature of the drug in the sample implies that the drug does not readily dissolve in $CDCl_3$. Under these circumstances, d_6-DMSO has been found to be a useful alternative solvent[4] that allows the identification of this group of drugs by NMR spectroscopy.

References

1. Available from: https://www.unodc.org/documents/scientific/barbiturates_and_benzodiazepines.pdf. Accessed 19 Aug 2024.
2. P. C. White, *J. Chromatogr.*, 1980, **200**, 271.
3. H. Lackner and G. Döring, *Arch. Toxicol.*, 1970, **26**, 220.
4. T. C. Kram, *J. Forensic Sci.*, 1978, **23**(3), 456.

10 The Analysis of Phenyl- and Benzylpiperazines

10.1 Introduction

Piperazines are a group of synthetic drugs that have similar effects in the human body to amphetamines and ring-substituted amphetamines. They have no direct analogues in nature. Piperazines and their precursors were first investigated as anthelmintic drugs in the 1950s, and later on, the piperazines themselves were trialled as antidepressants in the 1970s. Benzylpiperazine (BZP) abuse was first documented in the mid-1990s. This group of drugs is subject to different scheduling statuses across the globe, which makes the control of compounds very difficult. Piperazines came under control in Europe in 2008 after a risk assessment by the European Monitoring Centre for Drugs and Drug Addiction (EMCDDA). Controls are also in place in, for example, Australia, Canada, Japan, New Zealand, and the United States. These drugs have a reputation for being both "safe" and legal – neither of which is true. They have legitimate use in the synthesis of ciprofloxacin, quinoline, phenothiazines and some antidepressants.

Street samples of piperazines contain one or more benzylpiperazine and phenylpiperazine drugs (Figure 10.1). The drugs often occur in combination. Exemplar mixtures include one containing BZP mixed with trifluoromethyl piperazine (TFMPP) – a combination known as "A2". Other mixtures are more complex combinations of piperazines – for example, "X4", which is a mixture of *meta*-chlorophenylpiperazine (mCPP), *para*-chlorophenylpiperazine (pCPP), 3-trifluoromethylphenylpiperazine (TFMPP) and *ortho*-chlorophenylpiperazine (oCPP).[1] Piperazines may also be mixed with a range of other types of drugs,

---

Controlled Drug Analysis
By Michael D. Cole, Lata Gautam and Agatha Grela
© Michael D. Cole, Lata Gautam and Agatha Grela 2025
Published by the Royal Society of Chemistry, www.rsc.org

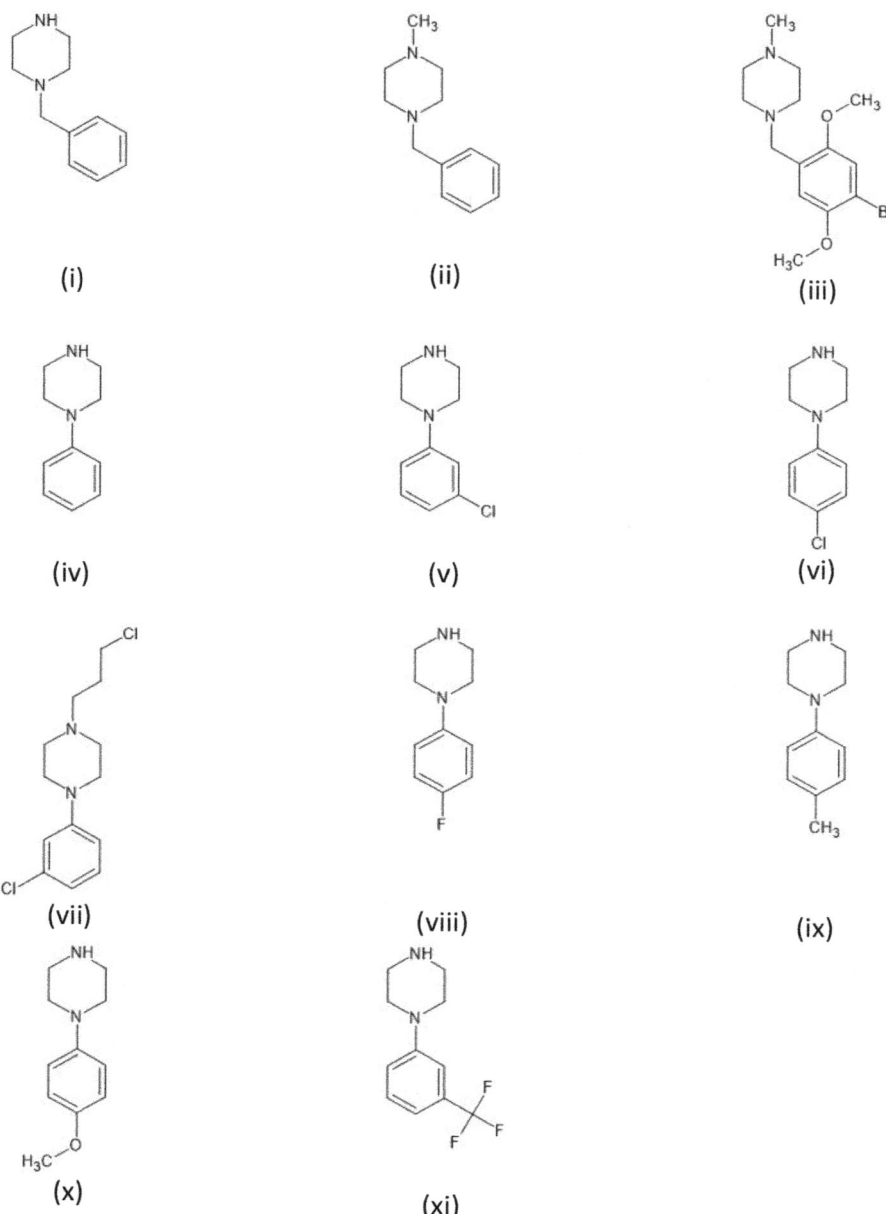

Figure 10.1 Structure of (i) 1-benzylpiperazine, (ii) 1-benzyl-4-methylpiperazine, (iii) 1-(4-bromo-2,5-dimethoxybenzyl)piperazine, (iv) 1-phenylpiperazine, (v) 1-(3-chlorophenyl)piperazine, (vi) 1-(4-chlorophenyl)piperazine, (vii) 1-(3-chlorophenyl)-4-(3-chloropropyl)piperazine, (viii) 1-(4-fluorophenyl)piperazine, (ix) 4-methylphenylpiperazine, (x) 1-(4-methoxyphenyl)piperazine and (xi) 1-(3-trifluoromethylphenyl)piperazine.

including caffeine, cocaine, dextromethorphan, diazepam, ephedrine, MDMA, methamphetamine and nicotinamide. BZP is often accompanied by the reaction by-product dibenzylpiperazine (DBZP), which is formed during the reaction process by which BZP is made. Its occurrence is due to the reaction of BZP with one of the starting materials, benzyl chloride, and indicates the route of manufacture.

Piperazines are most frequently found in tablet forms (Table 10.1). Dosages vary with drugs, with typical values being between 75 and 100 mg (BZP), 15 and 150 mg (mCPP) and 25 and 100 mg (TFMPP). Onset times may be short (mCPP, 20–60 min) up to more than 2 h (BZP).[2] The effects can last for long periods. For example, the effects of BZP can be felt 10–12 h after taking the drug. Some tablets contain considerable amounts of caffeine to mask the lack of the presence of the drug that the user is seeking.

Table 10.1 Examples of piperazines found in tablet dose forms illustrating exemplar compositions from samples of street drugs.

	Drugs identified in the tablet	Mass of drug (free base) in the tablet (mg)	Percentage of the mass of the tablet
	BZP	56	17
	TFMPP	106	32
	DBZP	8	2
			(Tablet = 333 mg)
	BZP	17	8
	TFMPP	21	9
	DBZP	1	0.5
			(Tablet = 224 mg)
	Ephedrine	36	10
	TFMPP	6	2
	Caffeine	101	28
			(Tablet = 355 mg)

10.2 Synthesis of Piperazines

Piperazines are relatively easily synthesised in straightforward one-pot reactions. For example, the manufacture of BZP can be as simple as dissolving piperazine hexahydrate in absolute alcohol, adding benzyl chloride, and shaking at 65 °C for 25 min, after which white crystals of BZP are formed. These crystals can be further cleaned up by washing with chilled ethanol and recrystallisation. Hydrochloride salt is formed by reacting BZP with hydrochloric acid. In the reaction mixture, BZP can react with the starting material benzoyl chloride and the reaction by-product dibenzyl-BZP (DBZP) is formed. Although not as active as BZP, this drug is also pharmacologically active and serves as a marker for the synthetic route used to make BZP.

Fluorinated phenylpiperazines are similarly easy to manufacture. Fluorinated anilines can be reacted with bis(2-chloroethyl)amines in the presence of a weak base while being heated, as shown in the synthesis of 1-(4-fluorophenyl)piperazine (4-FPP) (Figure 10.2).

A wide range of piperazines can be synthesised using a Pd-catalysed reaction protocol (Figure 10.3). By the appropriate choice of the starting materials, the piperazines illustrated in Figure 10.1 and the variants can be easily synthesised.

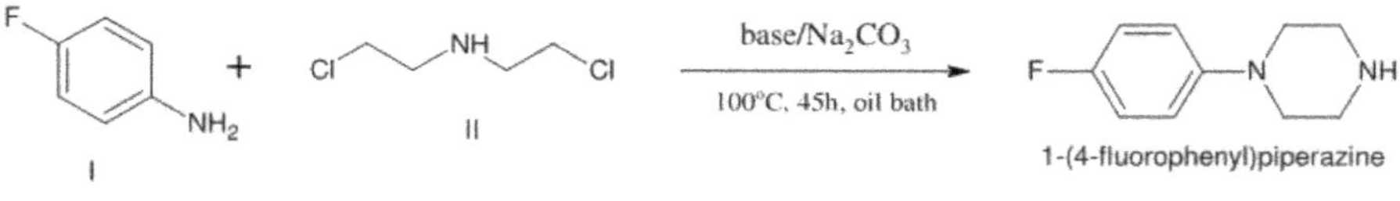

I = 4-fluoroaniline

II = Bis(2-chloroethyl)amine

Figure 10.2 Example of a synthetic route for the synthesis of 1-(4-fluorophenyl)piperazine.

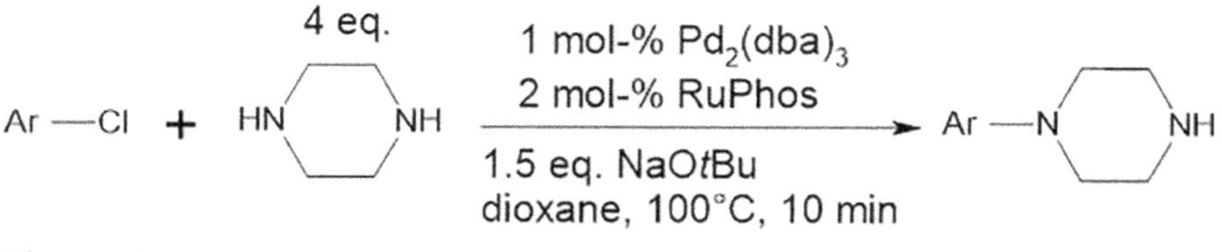

Figure 10.3 Synthesis of piperazines using a palladium catalyst protocol.

10.3 The Forensic Examination of Piperazines

The forensic examination of piperazines follows a typical route for bulk samples of drugs. Occasionally, a trace sample may be received – for example, material recovered from the inside of a die of a tabletting machine. In such cases, the analysis should proceed straight to an instrumental method. For bulk samples, presumptive and/or microcrystal tests might be used, but this will not identify which of the piperazines is/are present. In order to achieve this, instrumental methods must be applied. Thin layer chromatography is a useful adjunct method as a screening technique when comparing piperazine samples and avoids the need for an expensive instrumental technique to be applied, especially when an exclusion is achieved. Obviously, quantitative instrumental methods must be used to achieve a fully confirmed match between two or more tablets or seizures.

10.3.1 Physical Description of Piperazine-based Drugs

These drugs are generally encountered in two forms: powders and tablets. Powders should be treated as described in Chapter 2. If the seizure is of tabletted doses, then they should be sorted into visually identical groups and samples taken for analysis, again as described in Chapter 2. Great care should be taken not to damage the tablets in case there are ballistics that may be later useful in drug comparison. Some of the tablets may be coloured, and the analyst may be required to measure the colour using a microspectrophotometer under standard lighting conditions. This may add information that can be used when interrogating databases for comparative purposes to determine whether any related drugs have been seized.

10.3.2 Screening Tests for Piperazines

Presumptive tests should be carried out on material from the bulk of tablet dose forms that represents the bulk of the tablet or on a thoroughly homogenised powder sample. If examining a tablet, the material to be tested should be removed from the tablet well away from any "ballistics" features, identifying marks or logos that have been impressed on the tablet as each of these provides additional information when comparing samples. There are two types of presumptive tests that the analyst may use to identify parent compounds in a drug sample – microcrystal tests and colour tests. The second group of tests, the colour tests, find much wider applications than microcrystal tests and can be used to discriminate between primary and secondary amphetamines, ring-substituted amphetamines, and piperazines.

10.3.2.1 Microcrystal Tests for Piperazines

A very simple microcrystal test is available for testing BZP.[3] A small amount of the drug sample is reacted with aqueous mercury chloride (10 mg l^{-1}), and any crystals are allowed to nucleate and form. If BZP is present, small rectangular plates with birefringent edges will be observed. The size of the crystals is reported to be concentration-dependent. As the amount of BZP in the sample decreases, the crystals that form are larger, but fewer in number. The same test does not give a positive result with TFMPP, which frequently occurs in street drug samples containing BZP.

10.3.2.2 Presumptive Tests for Piperazines

Using combinations of common colour tests, namely Simon's reagent [comprising two solutions: the first is a mixture of 2% (w/v) sodium nitroprusside and 2% (w/v) acetaldehyde in water and the second is a solution of 2% (w/v) sodium carbonate in water], the Marquis reagent (100 ml of concentrated sulfuric acid mixed with 5 ml of 40% formaldehyde), and the 1,2-naphthoquinone-4-sulfonic acid sodium salt (NQS) reagent (5 drops of 0.1 M $NaHCO_3$–NaOH buffer solution, ensuring adequate mixing for a few seconds followed by the addition of 4 drops of 2.0×10^{-3} M NQS solution),[4] it is possible to discriminate between piperazines and drug classes with similar effects (Table 10.2). However, within each of these classes, it is not possible to identify which member or members of the class may be present, and as a consequence, instrumental techniques are required to carry out identification and quantitative analysis.

10.3.3 Dissolving Piperazines for Analysis

A large number of solvents have been suggested for the dissolution of piperazines. These include water, acetone, chloroform, diethyl ether and hexane.[5] The problem with these solvents is that at least one of the piperazines is not soluble in each of them. Additionally, compounds that have been mixed with piperazines, such as cocaine, may decompose in those containing water or may not be soluble in

Table 10.2 Reaction of piperazines, ring-substituted amphetamines and amphetamines with Simon's, Marquis and NQS reagents.

Reagent	Piperazines	1° ring-substituted amphetamines	2° ring-substituted amphetamines	1° amphetamines	2° amphetamines
Simon's reagent	Blue	No reaction	Blue	No reaction	No reaction
Marquis reagent	No reaction	Purple or blue	Purple or blue	Yellow or orange	Yellow or orange
NQS reagent	Red	Orange	Pink or orange	Orange or yellow	Pink or orange

the less polar solvents. One of the solvents that can be used is a low-molecular-weight ether – methyl *tert*-butyl ether (MTBE), which has been used to dissolve the street drug A2, a mixture of BZP and TFMPP.[6] Conveniently, this solvent is sufficiently polar to dissolve the drugs, while at the same time, it does not react with them.

Acidified methanol has also been used to dissolve piperazines.[7] Under these conditions, hydrochloride salt is generated, which renders the molecules most soluble in more polar solvents and solvents available for high-performance liquid chromatography. However, the problem with this method is that some of the piperazines are acid-labile,[8] which is also why chloroform cannot be used as a solvent for their dissolution. This is because chloroform breaks down in light over time, forming HCl as one of the breakdown products. In general, it is best to use the low-molecular-weight ethers described above (for example, MTBE) or to consider the use of a low-molecular-weight tertiary alcohol, such as tertiary butanol, to dissolve the samples.

10.3.4 Thin layer Chromatography of Piperazines

There is no piperazine-specific thin layer chromatography system that has been described. TLC systems for the analysis of alkaloids include mobile phases of methanol/ammonia (100 : 1.5 by volume), cyclohexane/toluene/diethylamine (75 : 15 : 10 by volume), and chloroform/acetone (4 : 1 by volume) with silica gel as the stationary phase. Similarly, the spray reagents that are used to visualise the piperazines after examination of the chromatogram under UV light include potassium iodoplatinate (0.15 g of potassium chloroplatinate and 3 g of potassium iodide in 100 ml of dilute HCl) and Dragendorff reagent [5 ml of a solution of 1.7 g basic bismuth nitrate in 100 ml of water/acetic acid (4 : 1), 5 ml of 40 g potassium iodide in 100 ml of water, 20 ml of acetic acid and 70 ml of water]. Neither of these are specific for piperazines, and a wide variety of alkaloids will have similar retardation factor (R_f) values and reactivities with the spray reagents. It is for this reason that while TLC can be used for a quick comparison of piperazine samples, it is not suitable for the identification and quantification of this group of compounds. In order to achieve these, more sophisticated instrumental techniques are required.

10.3.5 Instrumental Methods for the Analysis of Piperazines

10.3.5.1 High-performance Liquid Chromatography of Piperazines

High-performance liquid chromatography (HPLC) methods have been applied to piperazines. Early work[9] indicated that the baseline of a number of isomers of several piperazines could not be achieved and that the mass spectra obtained from these isomers could not be used to discriminate between these compounds. This is a common problem with some positional isomers of drugs, and is this why GC–MS is used to resolve the problem – it offers far greater resolution than other chromatographic techniques.

However, it is claimed that HPLC can be used in the identification and quantification of BZP and CPP.[10] In the system described, it was possible to separate isomers of CPP and to differentiate them on the basis of diode array detection. However, due to the similarity of the retention time of 3-CPP (7.32 min) and 4-CPP (7.27 min) and their UV spectra in the mobile phase, it is likely that if there were a mixture of these compounds, a GC-based technique would be required for definitive identification and quantification. What this illustrates is that it is always necessary to test a method on local equipment with locally sourced reagents – there may be sufficient resolution in some systems but not in others because of, for example, manufacturing differences in the chromatographic column or reagents.

While HPLC methods have been applied to piperazine drugs, due to the lack of resolution of some of the isomers and also the presence of other drug classes, this is not the method of choice for this group of drugs. Drug chemists must therefore turn to GC-based methods for piperazine identification and quantification.

10.3.5.2 Gas Chromatography–Mass Spectrometry of Piperazines

Early work on the GC–MS analysis of piperazines suggested that the retention times of different isomers of piperazines would be similar under typical GC conditions.[6] Additionally, the mass spectra of underivatised piperazines were virtually identical, as exemplified by the isomers of fluorophenylpiperazine (FPP) (Figure 10.4).

In order to address this issue, both acetylation and trifluoroacetylation of piperazines have been attempted.[6] Acetylation led to better

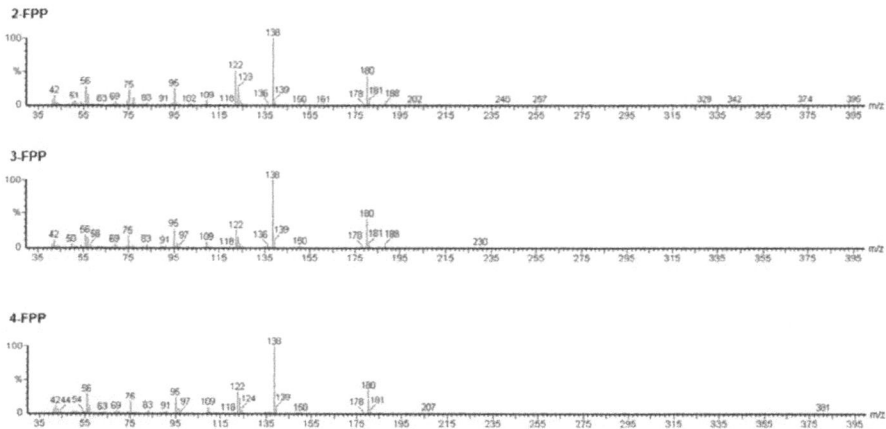

Figure 10.4 Mass spectra of the positional isomers 2-FPP, 3-FPP and 4-FPP[8] showing their near identical nature, illustrating that the identification of these compounds on the basis of the mass spectrum alone would be very difficult.[8] Reproduced from ref. 8 with permission from the author.

chromatography, but the mass spectra of the isomers were still not sufficiently differentiated. Trifluoroacetylation addresses this latter issue, but the derivative of *m*MeOPP could not be achieved. To achieve trifluoroacetylation, 50 µl of trifluoroacetic anhydride and 50 µl of ethyl acetate were added to the dried sample and incubated at 65 °C for 30 min. The mixture was blown dry under a gentle stream of dry nitrogen, redissolved in ethyl acetate, and subsequently analysed.

More recently, a study has been conducted on the piperazines and their congeners.[11] The drugs were analysed by GC–MS without derivatisation, thus avoiding the risk of contamination and incomplete derivatisation and reducing the cost and time required for the analysis. This did require, however, that the instrument was scrupulously clean. Samples were prepared in 2-methyl-propan-2-ol (*t*-BuOH) in which it was found that the piperazines were stable.[8] Using an instrumental setup with a BP-5 capillary column (30 m × 0.25 mm i.d. × 0.25 µm layer thickness), an injection temperature of 260 °C, an injection volume of 1 µl, a split ratio of 20:1, with He as the carrier gas at a flow rate of 1 ml min^{-1} and an oven temperature programme of 60 °C (1 min), rising at 10 °C min^{-1} to 170 °C (2 min) then to 280 °C (4 min) at 15 °C min^{-1}, transfer line and source temperatures of 280 °C, an ionisation energy of 70 eV, with the mass spectrometer in full scan mode, it was possible to separate

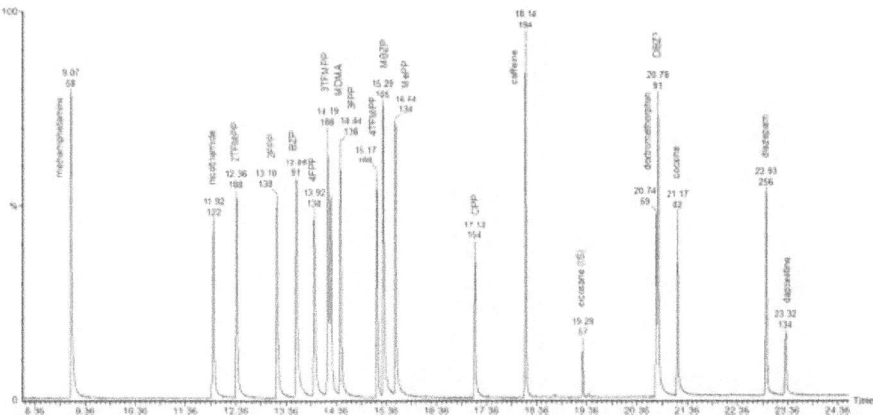

Figure 10.5 Chromatogram illustrating the gas chromatographic separation of piperazines and congeners illustrating that it is also possible to separate positional isomers.[8] Reproduced from ref. 8 with permission from the author.

and identify the piperazines and their congeners (Figure 10.5). It was also clearly possible to separate the isomers of, for example, FPP, which have similar mass spectra. On this basis, a positive identification could be made for these drugs. Using this method, it was also possible to identify dibenzylpiperazine. This impurity in the street sample indicates that the synthetic route to BZP was the reaction of piperazine with benzyl chloride, and it arises from the reaction of the benzyl chloride with the desired product, BZP, itself.

10.3.5.3 Gas Chromatography–Infrared Detection of Piperazines

Early work on the infrared spectroscopy of piperazines indicated that they could be identified using the (i) KBr disc and (ii) thin film methods.[9] However, in both cases, if there were mixtures of drugs or an infrared-active matrix as would be found in a tablet, this method would be problematic for the forensic drugs chemist.

With modern instrumentation, however, this logic can be extended, with GC separating the components of the mixture and the infrared detector being used as a confirmatory analytical technique. GC–IRD has been applied to the piperazines.[12] In some legislative systems, it is necessary to confirm the isomer of the piperazine being discussed and GC–IRD provides a useful tool in such analyses. Using data from the absorption maxima in the fingerprint region and just beyond (700 cm^{-1} to 1620 cm^{-1}), it is possible to discriminate the isomers of TFMPP, exploiting the fact that individual molecules will have unique spectra in the fingerprint region of the IR spectrum.

10.3.5.4 Electrochemical Methods for the Analysis of Piperazines

Electrochemical methods have been successfully applied to the analysis of piperazines.[1] Using cyclic voltammetry, it has proven possible to analyse a number of piperazines alone, for example, mCPP, in a blend (with MDMA) and in complex mixtures such as "X4" – a mixture of mCPP, TFMPP, oCPP and pCPP. The choice of electrode is important in drug chemistry. Carbon electrodes were found to foul, while boron-dipped diamond electrodes, used in aqueous solutions, were found to give a stable current (important for quantification) and to be mechanically robust and suffered from little fouling. By varying the buffers and surfactants used [including Triton X, sodium dodecyl sulfate (SDS) and cetyl pyridinium bromide hydrate (CPB)], it is also possible to overcome the effect of the adulterants and diluents, including aspirin, caffeine, cocaine, phenacetin, lidocaine, acetaminophen and MDMA, which are added to tabletted samples of piperazines. These characteristics, along with the use of small amounts of liquids and the relatively small size of the apparatus, offer the opportunity for the development of a field instrument for the determination of controlled substances.

References

1. L. R. Rocha, J. de Cássica Mendonça, T. B. Capelari, R. A. Medeiros and C. R. T. Tarley, *Sens. Actuators, B*, 2020, **310**, 127812.
2. C. D. Rosenbaum, S. P. Carreiro and K. M. Babu, *J. Med. Toxicol.*, 2012, **8**(1), 15.
3. M. Baron, M. Elie and L. Elie, *Drug Test. Anal.*, 2011, **3**(9), 576.
4. M. Philp, R. Shimmon, N. Stojanovska, M. Tahtouh and S. Fu, *Anal. Methods*, 2013, **5**(20), 5402.
5. https://syntheticdrugs.unodc.org/uploads/syntheticdrugs/res/library/forensics_html/Recommended_methods_for_the_identification_and_analysis_of_Piperazines_in_Seized_Materials_ST-NAR-47.pdf. Accessed 19 Aug 2024.
6. D. de Boer, I. J. Bosman, E. Hidvégi, C. Manzoni, A. A. Benkö, L. J. dos Reysand R. A. Maes, *Forensic Sci. Int.*, 2001, **121**(1–2), 47.
7. M. Takahashi, M. Nagashima, J. Suzuki, T. Seto, I. Yasuda and T. Yoshida, *Talanta*, 2009, **77**(4), 1245.
8. C. Kuleya, *The synthesis, analysis and characterisation of piperazine based drugs*, Doctoral dissertation, Anglia Ruskin University, 2014.
9. H. Inoue, Y. T. Iwata, T. Kanamori, H. Miyaguchi, K. Tsujikawa, K. Kuwayama, H. Tsutsumi, M. Katagi, H. Tsuchihashi and T. Kishi, *Jpn. J. Sci. Technol. Ident.*, 2004, **9**(2), 165.

10. S. Elliott and C. Smith, *J. Anal. Toxicol.*, 2008, **32**(2), 172.
11. C. Kuleya, S. Hall, L. Gautam and M. D. Cole, *Anal. Methods*, 2014, **6**(1), 156.
12. H. M. Maher, T. Awad and C. R. Clark, *Forensic Sci. Int.*, 2009, **188**(1–3), 31.

11 The Analysis of Cocaine

11.1 Introduction

Cocaine and related compounds are tropane alkaloids (Figure 11.1) derived from the coca plant, *Erythroxylon coca,* where cocaine occurs at concentrations of around 1% of the dry weight of the leaves. Historically, within the medical context, cocaine has been used as a local anaesthetic, particularly for the ear, nose and throat. Cocaine, when used as a recreational drug, is a stimulant that makes the user feel euphoric and talkative and increases the sense of mental alertness. Cocaine also increases sensitivity to sight, sound, and touch. The drug may temporarily act as an appetite suppressant and decrease the need for sleep. Less desirable effects include causing unpredictable violent or aggressive behaviour. In some South American cultures, coca leaves are chewed or brewed as a tea and used as an appetite suppressant and as a stimulant. Coca leaves are also used as flavouring agents.

Cocaine is generally encountered in one of two different forms, (i) crack cocaine and (ii) cocaine hydrochloride. Cocaine hydrochloride salt is a white powder and is freely soluble in water. Crack cocaine, the free base form of cocaine, is usually encountered in the form of small lumps. The amount of cocaine used ranges from 250 mg in a recreational user to several grams in addicts with a high tolerance to the drug. Cocaine hydrochloride is usually insufflated (also known as snorting, sniffing or blowing) or injected into veins. Crack cocaine is smoked alone or mixed with tobacco and/or cannabis.

In terms of control, the UN Single Convention 1961 classified cocaine in Schedule I, which lists the most harmful drugs with an associated high risk of abuse and where the strictest control possible

---

Controlled Drug Analysis
By Michael D. Cole, Lata Gautam and Agatha Grela
© Michael D. Cole, Lata Gautam and Agatha Grela 2025
Published by the Royal Society of Chemistry, www.rsc.org

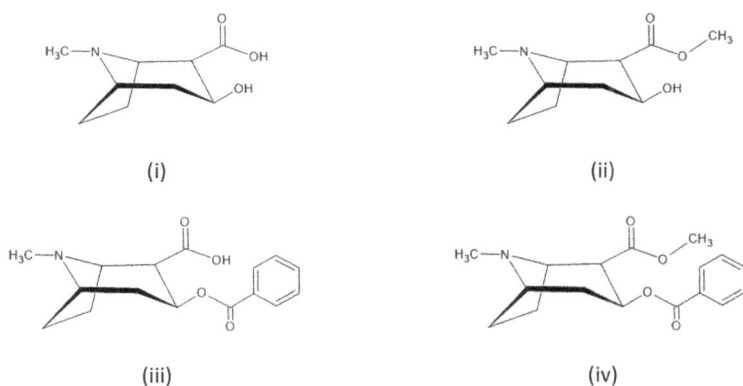

Figure 11.1 Structure of (i) ecgonine, (ii) ecgonine methyl ester, (iii) benzoyl ecgonine and (iv) cocaine.

is recommended. Cocaine solutions of concentrations lower than 0.1% are classified in Schedule III, where drugs are less strictly controlled as they are viewed as not leading to a risk of abuse. In the UK, cocaine is classified as a Class A drug under the Misuse of Drugs Act 1971. In both sets of regulations, coca leaves are also placed in the highest class of control, although the UN Convention does make provisions for the cultivation of coca bushes but only under strict governmental oversight. In Australia, cocaine is controlled under Section 8 of the Poisons Standard, and in the United States, it is a Schedule II drug.

As with most street samples of drugs, cocaine samples received in the lab for analysis are rarely pure. Cocaine is often contaminated with lidocaine and procaine (both local anaesthetics), sugars, starch, caffeine and phenacetin. The materials with which cocaine is mixed may depend on the region, and local knowledge may well help in expediting analysis. For example, in parts of the United States, cocaine is frequently mixed with sugars, while in the UK, local anaesthetics are sometimes used. The addition of these adulterants and diluents might serve different purposes, from increasing the mass of the sample (and therefore increasing profits of dealers) to being an attempt to hide the low concentration of the drug. Cocaine is also mixed with other drugs. For example, it is mixed with heroin where the taking of the two drugs together is known as "speedballing". Speedballing is a generic term where a polydrug mixture of a stimulant (in this case, cocaine) with a depressant (in this example, heroin) is taken by the user. Cocaine may also be mixed with ketamine and MDMA (a mixture known as "ketamollicaine"), LSD and MDMA (known as "candy flip"), PCP, or tobacco and cannabis.

11.2 The Extraction and Synthesis of Cocaine from Plant Material

Cocaine base is extracted from coca leaf (the dried leaves of *E. coca* or other cocaine-containing species of *Erythroxylon*) through several steps relying on the relative solubilities of the free base and salt form of the drug in immiscible nonpolar and polar (aqueous) solvents. The cocaine base is often subsequently oxidized with potassium permanganate to remove alkaloids that are co-extracted with cocaine. Cocaine base is then converted to cocaine hydrochloride via the use of hydrochloric acid. Crack cocaine is formed when cocaine hydrochloride is dropped into a solution of a base – sodium bicarbonate, for example. Crack cocaine, being relatively insoluble in aqueous solution, precipitates. The term "crack" comes from the noise made when this form of the drug is heated.

Cocaine is prepared in batchwise processes, and each step traditionally occurs on the multikilogram scale. Each batch has a unique chemical "signature" that comprises the purity of cocaine, the presence or absence of adulterants/diluents, total coca alkaloid content, cocaine hydrochloride conversion profile, and the cocaine base origin. It is rare to identify two batches that are exactly the same; there are typically differences in at least one part of the entire cocaine signature.

In order to monitor illicit drug production and availability in the market, some countries screen and control the supply chain of chemicals that are used to manufacture drugs. In the context of cocaine, these are

1. sulfuric acid used to dissolve coca paste in aqueous solution;
2. a permanganate salt (potassium or sodium) that can be mixed with the dissolved coca paste to help produce cocaine base;
3. ethyl ether, methyl ethyl ketone (MEK), methyl isobutyl ketone (MIBK), or toluene that can be used to dissolve the cocaine base; and
4. hydrochloric acid, used to prepare cocaine hydrochloride.

11.3 Analysis of Cocaine

11.3.1 Description and Sampling of Cocaine

Cocaine samples may be encountered in a variety of forms. Powdered seizures include large amounts of bulk powders seized through, for example, customs and excise operations through to street samples of individual doses. These may be wrapped in a wide variety of materials, including cling film "wraps", to balloons filled with individual doses of cocaine (Figure 11.2). When such seizures are examined, particular note should be taken of, for example, the knots used to tie the balloons. Often, particular distributors of drugs have a particular way to tie the knot. This can be used either as a "trade mark" of the particular distributor or to identify the content of the drug package.

Other materials that might be seized include items used to mix the drug or administer the drug. Under such circumstances, the item may be considered a trace rather than a bulk sample and should be treated accordingly. In all cases, a physical description of the item should be made, and decisions should be made regarding whether the material being examined is a bulk or a trace sample,

Figure 11.2 Examples of street samples of cocaine wrapped in uninflated balloons.

which should then be described, sampled (Chapter 2) and analysed accordingly.

11.3.2 Presumptive Tests for Cocaine

One of the most widely recognised tests for cocaine is the acidified cobalt thiocyanate test. The reagent is prepared as 2 ml of concentrated hydrochloric acid added to 98 ml of a 2% (w/v) solution of cobalt thiocyanate in water. A drop of the reagent is added to the test substrate, and if cocaine is present, a blue colour will develop. A modified version of this test is one where the test is carried out in a small test tube. The cobalt isothiocyanate is added to the test substrate. Chloroform is then added, and two layers form. If cocaine is present, the blue complex that is formed dissolves in the lower chloroform layer. The introduction of chloroform reduces the false positives occurring with the chemically related local anaesthetics procaine and lidocaine.[1] These molecules contain tertiary amines in their structures (Figure 11.3). However, procaine and lidocaine have additional primary and secondary amines, respectively. The increased polarity associated with these functional groups renders procaine and lidocaine more polar than cocaine, and therefore the drugs and coloured products are less soluble in chloroform.

Further colour tests include the Liebermann reagent [potassium nitrate in concentrated sulfuric acid (0.1 g ml^{-1})]. In the presence of cocaine, a yellow or orange colour forms depending on the amount of cocaine present. Cocaine also reacts with the Mandelin reagent (0.01 g of ammonium vanadate in 1 ml of concentrated sulfuric acid), forming an orange product.

11.3.3 Microcrystalline Tests for Cocaine

Microcrystalline tests for cocaine are another preliminary test recommended by the UNODC.[2] They are relatively quick and easy

Figure 11.3 Structure of (i) procaine and (ii) lidocaine.

to conduct and provide a cost-effective screen for some drugs. The assumption behind those tests is that each compound forms a unique (micro)crystal with a given reagent, which can be observed under a high power microscope or, preferably, a polarizing microscope. The crystals have specific shapes and cause specific rotation of light, which can be used to identify the drug in the sample under test.

The UNODC recommends two microcrystalline tests for cocaine examination, namely, the platinic chloride (H_2PtCl_6) and gold chloride ($HAuCl_4$) tests. For both tests, the sample should be at a concentration of 10 mg ml^{-1} dissolved in 10% hydrochloric acid. The reagents are prepared at 50 mg ml^{-1} in water. Two drops of the sample solution should be placed on a clean microscope slide, followed by the reagent being added close to the sample. The two solutions should not be directly mixed. A clean glass rod should be used to introduce the edges of the two solutions. As the solution dries, crystals are formed. A cover slide should not be used since the crystals that form may be deformed or crushed. Positive controls (to give crystal form, colour and rotation of light) and negative controls (to show that the reagent alone does not produce identical crystals) should also be completed.

11.3.4 Preparation of Solutions for the Analysis of Cocaine

Solvent choice is especially important in the analysis of cocaine, especially when samples are to be quantified or compared using chromatographic methods. This is because cocaine contains two ester groups and the related compounds ecgonine methyl ester and benzoylecgonine one each respectively (Figure 11.1). All three are therefore liable to hydrolysis in the presence of water. Additionally, any solvent used to prepare samples for analysis should conform to all of the requirements of a good solvent (freely dissolve the drugs, not form artefacts, not cause their breakdown, be compatible with subsequent analyses). It is for this reason that polar solvents such as low-molecular-weight alcohols (methanol, ethanol) should be avoided – because they contain water, which will accelerate the decomposition of these drugs. Conversely, solvents such as hexane are too nonpolar to dissolve the drugs quantitatively. On this basis chloroform is often chosen as the solvent for this drug group. It conforms to all of the requirements of a good solvent and at the same time being volatile is compatible with TLC and with GC or

GC–MS. However, it is not necessarily compatible with many HPLC mobile phases. HPLC is only rarely used to analyse cocaine, but where it is the mobile phase itself, or methanol, it should be used to prepare samples that should then be analysed as quickly as possible after preparation to reduce the risk of hydrolysis. Ahead of analysis, samples should be dissolved in the solvent of choice at a known concentration, ideally resulting in a cocaine concentration in the middle of a calibration curve where quantitative methods are being applied. A concentration of 100 µg ml^{-1} is a good starting point. Samples should also be filtered or centrifuged to remove any solid, insoluble material ahead of analysis.

11.3.5 Thin Layer Chromatography of Cocaine

There are a number of published TLC methods to analyse cocaine. The vast majority use normal phase chromatography systems with silica gel impregnated with a fluorescent indicator as the stationary phase. The excitation wavelength for the indicator is usually 254 nm, and on visualisation, cocaine and related compounds appear as purple spots on a bright (usually yellow) background. Such detection methods are not specific to cocaine because any compound that absorbs UV light at 254 nm will appear the same.

The UNODC-recommended method[2] uses such a stationary phase and a mobile phase of either (i) chloroform:dioxane:ethyl acetate:29% ammonia (25:60:10:5, by volume), (ii) methanol:29% ammonia (100:1.5, by volume) or (iii) cyclohexane:toluene:diethylamine (75:15:10, by volume). Cocaine hydrochloride will exhibit poor chromatographic properties because of the polar nature of the hydrochloride salt and the interactions with the silanol groups on the silica gel. In all three mobile phases, ion suppression is used, where the base (ammonia) reacts with the HCl, ensuring that the compounds of interest are in the free base form, which ensures better chromatography through reducing the tailing that occurs with polar compounds on silica gel. To improve the chromatography yet further, some analysts deactivate (dry out) the silica gel plate at 80 °C for 20 min prior to use. When running these systems, standards of cocaine and related drugs should also be analysed, with the solutions at the same order of magnitude of concentration as the suspected sample. Additionally, solvent blanks should also be included in each analysis to demonstrate that the solvents and apparatus are free of drug contamination. The chromatogram, once developed, is visualised using UV light (254 nm) and one of a number of spray reagents including potassium iodoplatinate (0.15 g of potassium chloroplatinate and 3 g of potassium iodide in 100 ml of dilute HCl) or Dragendorff reagent (5 ml of a solution of 1.7 g of basic bismuth nitrate in 100 ml of water/acetic acid (4 : 1), 5 ml of 40 g potassium iodide in 100 ml of water, 20 ml of acetic acid and 70 ml of water). Under UV light, cocaine and related compounds will appear as purple spots against a fluorescent background. With the potassium iodoplatinate reagent, the compounds are visualised as purple spots against a brown background and with the Dragendorff reagent as orange spots against a pale orange background.

11.3.6 Gas Chromatography–Mass Spectrometry of Cocaine

Due to the fact that cocaine is easily identified by GC–MS, it is the method of choice for the confirmatory analysis of drugs containing cocaine and related alkaloids. The drug mixtures can be analysed with the drugs in their native form or derivatised. The UNODC method described for analysis of native drugs[2] uses a capillary DB-5 column (or equivalent) with a temperature programme set initially at 60 °C, held for 3 min, with the temperature then raised at 40 °C min^{-1} to 300 °C, and held for 6 min. However, even with fused silica capillary columns, it is problematic to ensure that the column temperature is "keeping up" with the oven temperature when the temperature is raised at this rate. It is for this reason that other oven temperature programmes might be considered. A commonly employed system used an OV-1 column or equivalent (25 m × 0.25 mm i.d. × 0.5 μm), with an injection temperature of 300 °C and a column temperature programme of 170 °C (2 min) increased to 280 °C at 16 °C min^{-1} and held isothermally for 2 min. The carrier gas is typically helium at a flow rate of 1 ml min^{-1}, and the split ratio set appropriately for the drug concentration. Standard mass spectrometric conditions can be easily applied. Cocaine exhibits good chromatographic properties and a very characteristic mass spectrum (Figure 11.4). However, the breakdown products of cocaine, namely benzoyl ecgonine, ecgonine methyl ester and ecgonine (Figure 11.1), all contain polar functional groups and may exhibit tailing if the chromatographic system is not scrupulously clean. It is for this reason that derivatisation may be employed.

A number of derivatisation reactions may be employed for cocaine analysis. Typically, the samples, a positive control and solvent blank are prepared containing an internal standard. N-alkanes may be used, or 2,2,2-triphenylacetophenone (benzopinacolone), as recommended by the UNODC,[2] at a final concentration of 50 μg ml^{-1}.

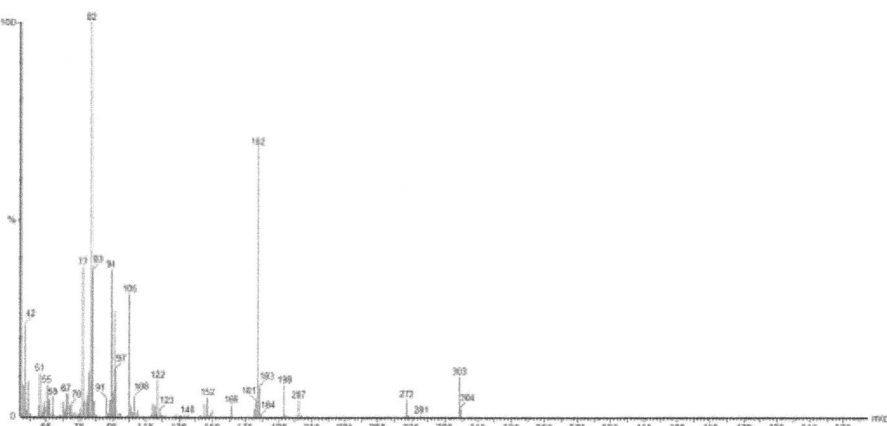

Figure 11.4 Mass spectrum of cocaine.[3] Reproduced from ref. 3 with permission from the author.

The mixtures are prepared in chloroform and then dried under a stream of dry nitrogen, used to reduce the risk of hydrolysis and oxidation of the alkaloids. It is to this mixture that the derivatisation reagent can be added. Reagents include 10% N,O-BSA in n-hexane, which will form the trimethylsilylethers as derivatives of the ecgonine compounds, or acid anhydrides, for example BSTFA and TCMS (99 : 1) which forms the trifluoroacetylated products. The derivatives are far less polar and hence exhibit better chromatographic properties. Derivatisation will also increase the sensitivity of the analytical method.

11.3.7 High-performance Liquid Chromatography of Cocaine

Cocaine is not routinely analysed using liquid chromatography. However, some HPLC systems have been described that use reversed-phase and/or ion exchange processes to separate the drugs contained in the mixture.

A reversed-phase system using an ion exchange system is the method recommended by the UNODC.[2] A 16 cm × 5 mm i.d. C-18 column, with a mobile phase of methanol : water : 1% phosphoric acid : n-hexylamine, 300 : 700 : 1000 : 14 by volume, pH 2.5, is recommended with UV detection at 230 nm. In this case, the hexylamine will partition into the ODS stationary phase and, because of the pH, be positively charged. The alkaloids will similarly be positively charged at this pH, and so in addition to the partition processes that occur on reversed phase stationary phases, ion exchange processes will occur, effecting separation of the alkaloids.

A similar principle has been applied using formic acid rather than phosphoric acid and hexylamine.[4] In this case, the system employed was an Acquity UPLC BEH C18 column (2.1 × 150 mm i.d., 1.7 μm particle size), with a mobile phase of A (0.5% formic acid : 99.5% acetonitrile) and B (0.5% formic acid) with the use of an elution gradient. The mobile phase gradient conditions were 10% A for 1.5 min, increasing to 60% A after 14 min and again increasing to 100% A in 15.5 min and holding for 7 min. Detection was by mass spectrometry. In this case, the presence of the formic acid and the low pH cause the residual silanol groups on the silica gel to ionise and the cocaine to be in the charged form. As a consequence, the separation is achieved by a partition and ion exchange mechanism. Interestingly, this method, although designed for the analysis of drugs in wastewater, could equally be applied to emerging polydrug mixtures containing cocaine.

11.3.8 Ultraviolet Spectroscopy of Cocaine

In some parts of the world, cocaine is mixed with sugars rather than other controlled and recreational drugs. Under such circumstances, UV spectroscopy can be used to easily quantify the amount of cocaine after its presence has been confirmed using, for example, qualitative GC–MS methods. In these cases, a series of dilutions of cocaine are prepared in methanol where the concentration of the drug is calculated as a concentration of free base. This is because any street sample of cocaine may potentially contain more than salt form, or the drug may be present as both free base and hydrochloride. In such cases, the amount of drug in the various salt forms will not be known and so it is best to work with the drug corrected for the free base form. Additionally, samples should be freshly prepared because methanol is hygroscopic and will absorb water, which in turn will lead to hydrolysis of the cocaine. Methanol is used because it is transparent to UV light in the wavelength range of 210–350 nm. A calibration curve of absorbance against the concentration of the drug as free base is constructed, and the regression equation is calculated. The absorbance of the sample at a known concentration is measured, and from the calibration curve, the concentration of the drug, as free base, is calculated. The wavelength used for calibration is usually the absorption maximum of the drug, in the case of cocaine calculated. The wavelength used for calibration is usually the absorption maximum of the drug, in the case of cocaine 232 nm because this provides the greatest sensitivity and longest linear dynamic range for the calibration curve. Interestingly, UV spectroscopy can be used for the tentative identification of cocaine through the UV spectrum obtained in methanol, as can the triplet that is obtained from the second derivative of the UV spectrum of cocaine in methanol at around 232 nm.[5] This is important because not all forensic science laboratories have power 24 h a day and hence running GC–MS instruments is not possible. The use of UV spectrophotometers, however, is possible, because once they are switched on they are usable once they are warmed up. It should be remembered, however, that their functionality must be demonstrated through the use of holmium filters prior to any casework analysis being undertaken. The use of the derivatives of spectroscopic data provides alternative means by which identification of drugs can be made.

11.3.9 Infrared Spectroscopy of Cocaine

In addition to the identification and discrimination of cocaine and cocaine hydrochloride using infrared spectroscopy,[2] it is possible to quantify cocaine in street samples of drugs where the drug (cocaine) is mixed with a simple mixture of, for example, sugars. In such cases, a calibration curve can be prepared from the IR absorbance ratio of the absorbance due to the hydroxide group in the sugar at around 3600 cm^{-1} relative to that due to the carbonyl group of cocaine (~1728 cm^{-1}) against the percentage of cocaine free base in the sample. As the concentration of cocaine increases, the ratio decreases. A regression equation can then be calculated, and from the experimental ratio obtained from the street sample, the concentration of cocaine in the sample can be calculated. Such methodologies do make a number of assumptions. These include the assumption that the sample only consists of cocaine and sugar where the infrared absorptions of the two materials are independent of each other. Another assumption is that the drug is not chemically decomposing into other compounds that can absorb IR radiation at the same wavenumber.

11.3.10 Analysis of DNA in Cocaine

There is little work on the analysis of DNA in cocaine. Some studies have focused on herbal material of different species of *Erythroxylon*, in order to establish their phylogenetic relationships and/or their origin, rather than identify them as material from this genus *per se*. Using the analysis of amplified fragment length polymorphisms (AFLP), it has been possible to identify the geographic origin of four taxa of *Erythroxylon* which produce cocaine.[6] Using AFLP, it has proven to be possible to differentiate the different species and subspecies and also their geographic origin and whether the samples came from Colombia, Bolivia or Peru. While this work only describes the analysis of a small sample, it may complement the use of other techniques in determining the geographic origin of herbal *E. coca* samples. The work has been extended by to differentiate species of *Erythroxylon* from different regions on different continents (Africa, tropical Americas, and Asia Pacific),[7] but a great deal more work is required before such work finds application in forensic drug analysis. Studies have also shown that if this method is to find forensic science applications, in order to preserve the DNA for analysis, plant material should be stored at 4 °C or −20 °C, without air or oven drying.

11.3.11 Stable Isotope Analysis and Cocaine

There is forensic interest in the geographic source of powdered cocaine. While it may be possible to achieve this for plant material through the analysis of DNA and AFLP analysis (Section 11.3.10), the analysis of DNA from powdered cocaine is more problematic because the chemical processing of plant material to produce cocaine will result in the degradation of the DNA itself. An alternative strategy is to examine the stable isotope ratios found in cocaine, especially the carbon and nitrogen ratios.[8] Through such studies, it has been possible to authenticate the origin of cocaine coming from Bolivia, Peru, Ecuador and Colombia. However, such studies require that authenticated materials from each are available in order to determine the expected isotopic ratios, and this can be problematic. Additionally, isotopic fractionation during cocaine processing needs to be fully understood before the country of origin or geo-location of the cocaine source can be determined. While the carbon and oxygen ratios remain stable, the hydrogen and nitrogen ratios are liable to exhibit variation due to isotopic fractionation.[9] A comprehension of such chemical changes across the known sources of cocaine is required prior to using stable isotope analysis in determining geo-location and source of cocaine in the forensic context.

References

1. K. A. Kovar and M. Laudszun, in *Chemistry and reaction mechanisms of rapid tests for drugs of abuse and precursors chemicals*, SCITEC/6, United Nations, Vienna, 1989.
2. Available from: https://www.unodc.org/unodc/en/scientists/recommended-methods-for-the-identification-and-analysis-of-cocaine-in-seized-materials.html. Accessed 19 Aug 2024.
3. C. Kuleya, *The synthesis, analysis and characterisation of piperazine based drugs*, Doctoral dissertation, Anglia Ruskin Research Online (ARRO), 2014.
4. Y. Peng, S. Hall and L. Gautam, *TrAC, Trends Anal. Chem.*, 2016, **85**, 232.
5. A. A. Fasanmade, *Afr. J. Med. Med. Sci.*, 1994, **23**(4), 333.
6. E. L. Johnson, J. A. Saunders, S. Mischke, C. S. Helling and S. D. Emche, *Phytochemistry*, 2003, **64**(1), 187.
7. S. D. Emche, D. Zhang, M. B. Islam, B. A. Baileyand and L. W. Meinhardt, *Trop. Plant Biol.*, 2011, **4**, 126.
8. J. R. Ehleringer, D. A. Cooper, M. J. Lott and C. S. Cook, *Forensic Sci. Int.*, 1999, **106**(1), 27.
9. J. R. Mallette, J. F. Casale, L. M. Jones and D. R. Morello, *Forensic Sci. Int.*, 2017, **270**, 255.

12 The Analysis of Benzodiazepines

12.1 Introduction

In addition to dealing with illicitly synthesised drugs of abuse, forensic scientists encounter challenges related to stolen and counterfeit pharmaceuticals and designer variations of such drugs. One such group of drugs is benzodiazepines, which include both traditional benzodiazepines and those regarded as "designer" benzodiazepines. Designer benzodiazepines are formed by subtle chemical modifications of prescription benzodiazepines; for example, the introduction of chlorine to diazepam yields diclazepam. Other benzodiazepines that find their way onto the recreational drug market have failed clinical trials. Examples include diclazepam, pyrazolam, and flubromazepam[1] (Figure 12.1). As pharmaceuticals, benzodiazepines are central nervous system depressants introduced for medical use in the 1960s. Initially replacing the more toxic barbiturates, they were prescribed for treating anxiety, sleep disorders, as sedatives, anticonvulsants and to treat the symptoms of alcohol withdrawal. However, concerns have arisen due to overprescription, ease of availability, and the potential for misuse and dependence associated with traditional benzodiazepines.

Benzodiazepines can be categorised into six distinct substructural groups, with the 1,4-benzodiazepine core being the most predominant, constituting the majority of benzodiazepines in circulation (*e.g.* diazepam, flunitrazepam, nitrazepam and temazepam) (Table 12.1). The remaining five substructures include oxazolobenzodiazepines (*e.g.* ketazolam and oxazolam),

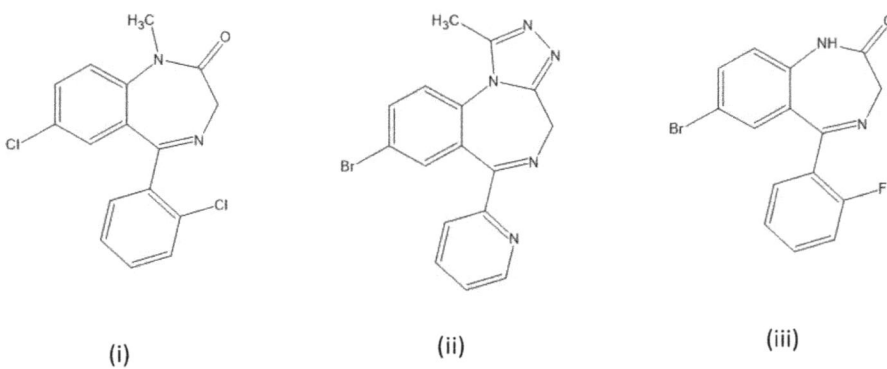

Figure 12.1 Examples of benzodiazepines that have been found in the illicit drug market following failure in clinical trials: (i) diclazepam, (ii) pyrazolam and (iii) flubromazepam.

thienotriazolodiazepines (*e.g.* brotizolam and etizolam), triazolobenzodiazepines (*e.g.* clonazolam, flualprazolam, flubromazolam, and triazolam), imidazolobenzodiazepines (*e.g.* loprazolam), and thienodiazepines (bentazepam and clotiazepam).

Benzodiazepines are used as recreational drugs and may be taken singly or they may also be found in drug mixtures, for example, in heroin, in Happy water – a mixture that may contain methamphetamine, diazepam, caffeine and tramadol – or Fairy water, which is a mixture of codeine, MDMA and nimetazepam. These drugs, especially flunitrazepam (Rohypnol), are also detected in cases of drink spiking and drug-facilitated sexual assault.[2]

Benzodiazepines are mostly encountered, in the forensic science context, as capsules and tablets. Those involved in drug control should also be aware that they also occur as powders, injectable solutions, syrups and suppositories. Designer benzodiazepines are marketed under various terms such as new psychoactive substances (NPS), legal highs, and research chemicals. Both traditional and designer benzodiazepines have a wide variety of street names, including benzos, downers, nerve pills, blues/blueys, tranx, tranks, roche's, mother's little helpers, duck eggs (temazepam), roofies (Rohypnol) and V's. Pharmaceutical forms are available as carefully controlled doses, *e.g.* of the order of 2.5–10 mg for benzodiazepines and taken as prescribed. However, the dose taken by recreational users will depend on the desired effect, among other factors. Doses of common designer benzodiazepines (0.08–30 mg) have a much wider range than those of pharmaceutical forms and, in some cases, are much lower than the traditional benzodiazepines on which they

Table 12.1 Exemplar benzodiazepines representing the six structural classes on this group of drugs (i) diazepam, (ii) flunitrazepam, (iii) ketazolam, (iv) oxazolam, (v) brotizolam, (vi) etizolam, (vii) clonazolam, (viii) flualprazolam, (ix) loprazolam, (x) bentazepam and (xi) clotiazepam.

Benzodiazepine class	Exemplar benzodiazepines	
1,4-Benzodiazepine core structure	(i)	(ii)
Oxazolobenzodiazepines	(iii)	(iv)
Thienotriazolodiazepines	(v)	(vi)

(continued)

Table 12.1 (continued)

Benzodiazepine class	Exemplar benzodiazepines	
Triazolobenzodiazepines	(vii)	(viii)
Imidazolobenzodiazepines	(ix)	
Thienodiazepines	(x)	(xi)

are based (Table 12.2). It is important to know these relative doses when setting up, for example, instrument calibration curves where the amount expected in a street sample is ideally located within the middle of the concentration range of the linear part of the instrument calibration curve.

Pharmaceutical benzodiazepines come from many countries, including Italy, India, China and Germany. Traditional

Table 12.2 Comparison of the doses of traditional benzodiazepines and their designer benzodiazepine counterparts showing the much lower doses required for the latter.

Benzodiazepine	Dose	Designer variant	Dose
Nitrazepam	5–10 mg	Nitrazolam	1.2 mg
Flubromazepam	4–8 mg	Flubromazolam	0.2–0.4 mg
Clonazepam	0.5 mg	Clonazolam	0.2–0.4 mg
Bromazepam	5 mg	Pyrazolam	2–3 mg

benzodiazepines are pharmaceutical compounds controlled internationally. At the time of writing this book, most of the designer benzodiazepines were sourced from China as bulk powders, processed into tablets in Europe and sold as replacements for prescribed benzodiazepines. However, a small number of benzodiazepines have also been sourced from India and fake diazepam tablets have been produced in Scotland, UK, from etizolam. The drug market and its source change constantly, influenced by many factors including the availability of precursors and production costs.

In terms of control, benzodiazepines were initially classified under Schedule IV of the 1971 United Nations Convention on Psychotropic Substances in 1984. However, in 1995, flunitrazepam was reclassified, being moved from Schedule IV to Schedule III. This decision was linked to flunitrazepam being cited as one of the most widely abused benzodiazepines and its frequent diversion into the illicit market. In the United Kingdom, benzodiazepines are controlled as Class C drugs, with each type listed individually under the Misuse of Drugs Regulations 1985. Temazepam and flunitrazepam are categorised under Schedule 3 control, under the Misuse of Drugs Regulations 2001, whereas other benzodiazepines are placed under Schedule 4. In the United States, benzodiazepines are classified as Schedule IV drugs.

12.2 Analysis of Benzodiazepines

Benzodiazepines present an ongoing analytical challenge to the drug chemist because of the continuing emergence of new members of this drug class. However, the standard scheme for sampling and drug analysis (Chapter 2) should be followed. According to the UNODC,[3] three distinct analytical techniques are recommended for drug detection: colour tests, chromatography (such as TLC, GC, or HPLC), and spectroscopy (for example, IR or UV). Notably, if GC–MS is employed, it is considered as two techniques, as long as the results from both methodologies (*i.e.* retention time and mass spectrum) are used.

12.2.1 Sampling and Physical Description of Benzodiazepines

Sampling is dependent on the form, number and type of dosage units seized, and guidance is provided in Chapter 2. Whatever the dose form and number of dose units seized, the critical point is to ensure that a representative sample of the whole seizure (generally multi-unit) is taken for analysis. The next step involves thoroughly examining the items and making a complete physical description of the materials seized, followed by a comparison with relevant references to determine if the drug originates from a legal source. One such example is the use of TICTAC, a specialised drug identification, drug analysis and drug information database. However, drug chemists should remember that while the dose units may look like those from legitimate sources, it does not mean that the drug contained within is that described by the database and a full definitive identification of the drug in the samples is still required. This is another general principle of drug analysis – the identity of the controlled substance(s) must always be proven. Visual and physical comparison with a database provides leads as to drug identity but does not prove it. However, if the substance is suspected to be from an illicit source (*i.e.* not resembling a recognised pharmaceutical product), a comprehensive chemical analysis must equally be conducted.

12.2.2 Presumptive Tests for Benzodiazepines

As benzodiazepines contain a wide range of functional groups, there is no specific colour test for this class of drug. However, the Zimmermann test provides a convenient screening method for this group of drugs. The Zimmermann reagent is prepared as two solutions. The first is a solution containing 2,4-dinitrobenzene (1% w/v) in methanol, and the second is a 15% KOH (aq) solution. The test is performed by adding one drop of the first reagent to the test substrate, followed by one drop of the second. The development of a red, purple or pink colour indicates the possible presence of benzodiazepines. Most benzodiazepines also give an orange colour with a variant of the Marquis test (5 ml of formaldehyde and 100 ml of concentrated sulfuric acid) when heated at 100 °C for 1 min. Flunitrazepam gives a pink colour. With the emergence of designer benzodiazepines, further research into colour test screening for this drug group will be required.

12.2.3 Dissolution of Benzodiazepines for Further Analysis

As with many other classes of drugs, benzodiazepines commonly exist in either the free base or salt forms. They exhibit solubility in chloroform, acetone, ethanol and ether. Consequently, for sample preparation and subsequent analysis, dissolving benzodiazepines in an appropriate solvent is the first step. The solvent should comply with all of the requirements of a good solvent for drug analysis, but with the occurrence of new designer benzodiazepines, the analyst should be aware of the potential for the drug to react with the solvent or to decompose in solution. This can be determined, in principle, by using "paper chemistry" and working out whether there might be a reaction. Practically, using "check standards" throughout any instrumental analytical sequence will help determine whether sample decomposition is occurring. If the response from the check standard diminishes, then the sample is probably decomposing in some way. Once the solution is prepared, should any undissolved particles be observed or the solution appear cloudy, filtration or centrifugation will be required. Evaporation of the solvent under a stream of dry nitrogen is then recommended, facilitating the reconstitution of the drug in an internal standard to enable accurate identification and quantification depending on the technique being employed.

12.2.4 Thin Layer Chromatography of Benzodiazepines

Following the completion of colour tests, the subsequent analytical step for benzodiazepines is TLC. Benzodiazepine standards, dissolved in an organic solvent (Section 12.2.3) along with the test sample, can be analysed for qualitative determination of benzodiazepines and adulterants. The retention factor is calculated for each compound/spot present in a sample, which is then compared with the standard to tentatively identify the compounds present in the test sample. Importantly, exclusions can also be achieved using this method – an important consideration when thinking about whether or not a drug might be one of the designer benzodiazepines. An exclusion using TLC is much more cost-effective than the one achieved using expensive instrumental analysis. Simultaneous analysis of samples, standards, and negative controls on the TLC plate is recommended.

As this is a diverse group of drugs, the use of three TLC mobile phases in combination is recommended[3] followed by visualisation of the separated compounds. Separation is achieved on a normal-phase silica gel TLC plate impregnated with a fluorescent indicator, usually one that fluoresces at 254 nm. This means that under UV light at 254 nm, the benzodiazepines appear as purple spots on a light background. Three mobile phases can be used: (i) chloroform/acetone (80 : 20 by volume), (ii) chloroform/methanol (90 : 10 by volume), or (iii) cyclohexane/toluene/diethylamine (75 : 15 : 10 by volume). When using the latter system, plates should be dried at 120 °C for 5 min to remove diethylamine from the plate. Alternatively, a hair dryer (hot air blower) can be used. This is one of the rare occasions that it is acceptable to heat a TLC chromatogram after development because, in this case, the benzodiazepines, with relatively high molecular weights, are not evaporated in the warm air stream. When developed, the chromatogram should be observed under UV light at 254 nm followed by spraying with 1M H_2SO_4 followed by heating at 80 °C for 5 min. Any visualised compounds appear as brown spots on a yellow/light brown background. However, a wide range of organic compounds react in this system, which is not specific for benzodiazepines – the sulfuric acid acts as a dehydrating agent, and the brown colour occurs because of the dehydration of organic molecules. Alternatively, the chromatogram can be sprayed with an acidified potassium iodoplatinate reagent (0.15 g of potassium chloroplatinate and 3 g of potassium iodide in 100 ml of dilute HCl), which results in purple-coloured spots on a brown background. Again, this reagent is not specific to benzodiazepines and reacts with a wide range of alkaloidal compounds.

12.2.5 Gas Chromatography-based Analysis of Benzodiazepines

Benzodiazepines have been analysed using gas chromatography with a variety of detection methods. However, there are issues of thermal degradation and rearrangement reactions occurring in some members of this drug class, particularly with the 3-hydroxy derivatives. The problem is exacerbated by the decomposition of one benzodiazepine into another with the concomitant risk of misidentification. By way of example, ketazolam decomposes into diazepam. The use of mass spectrometry adds value in terms of identifying and confirming the presence of these compounds. However, HPLC is the technique of choice for the analysis of benzodiazepines.

If benzodiazepines are to be analysed by GC-based techniques, derivatisation is required for some benzodiazepines (*e.g.* temazepam, oxazepam, and nitrazepam) (Figure 12.2), adding an additional step to the sample preparation process. Derivatisation using silylation methods (*e.g.* BSTFA or BSTFA:TMCS or TBDMS) is commonly employed. However, these reagents are expensive and toxic. Sometimes, after derivatisation, unwanted peaks are also detected in the chromatogram, thereby increasing data analysis and result interpretation time.

As with any other drug analysis using GC-based techniques, the addition of an internal standard at the beginning of the sample preparation is a very important step. This allows compensation for variation in sample preparation and injection

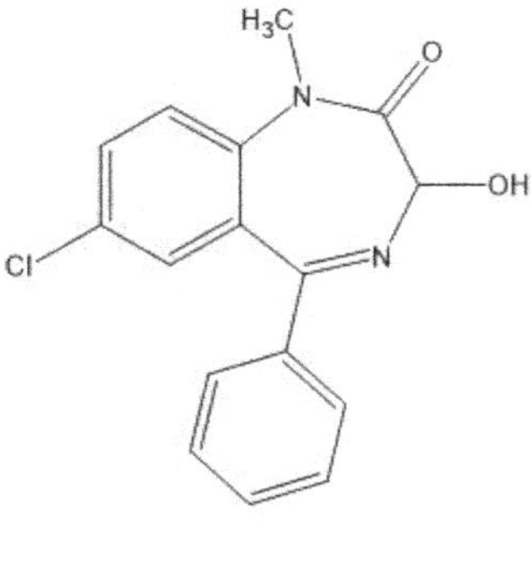

Figure 12.2 Examples of benzodiazepines that contain polar functional groups that require derivatisation before analysis by GC–MS: (i) temazepam, (ii) oxazepam, and (iii) nitrazepam.

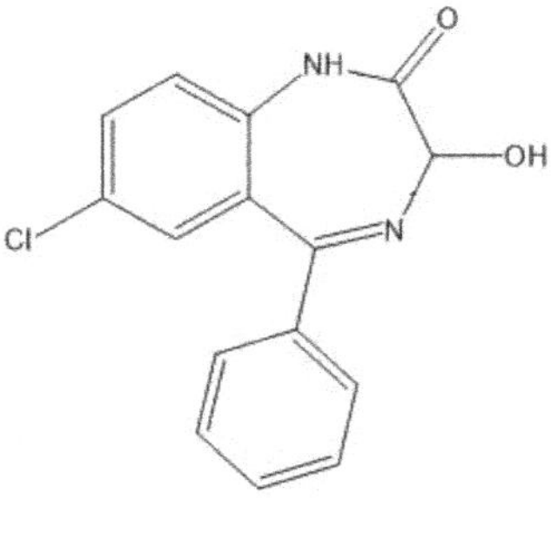

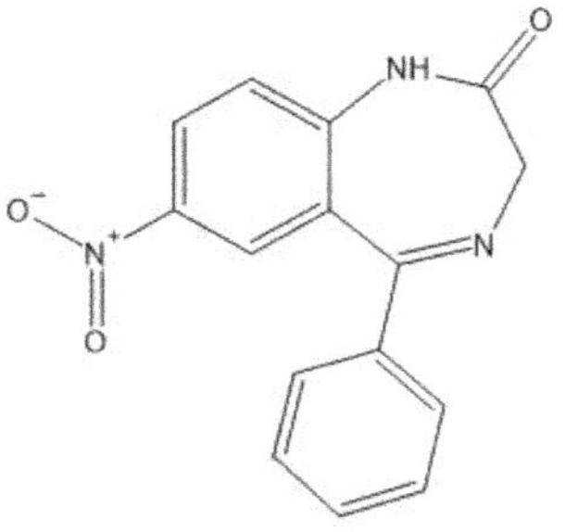

variability. Both isothermal and temperature gradient systems can be used (Table 12.3).

Identification of benzodiazepines can be made using GC retention indices. However, it is important to run the standards using the same method on the same instrument as part of the same analytical sequence, and hence, a better comparison can be made. Due to the complexity of the nature of this group of compounds and the risk of thermal decomposition when carrying out gas chromatographic analysis, chromatograms should be inspected for their quality, background noise and peak properties. Retention time data should be collected followed by the calculation of the retention factor and then the peak integration for quantification. Peak area ratios (*i.e.* peak area of standard/peak area of the internal standard) should be used for quantification. Initial identification of the

Table 12.3 GC–FID and GC–MS conditions and parameters used for the analysis of benzodiazepines.

	GC–FID parameters	GC–MS parameters[3]
Oven programme	275 °C, isothermal	135 °C (4 min), increased to 200 °C at 13 °C min^{-1}, increased to 312 °C at 6 °C min^{-1} and hold for 6 min
Column details	BP-1, 25 m length × 0.25 mm i.d. 0.25 µm film thickness	BP-1 or DB-1, 10–15 m length × 0.32 or 0.53 mm i.d. 1.5–3 µm film thickness
Carrier gas	He, 1 ml min^{-1}	He, 1 ml min^{-1}
Injector	20 : 1, 275 °C	Split/splitless
Injection port	250 °C	250–300 °C
Split ratio	20 : 1	Split/splitless
Detector	275 °C, hydrogen 35 ml min^{-1}, air 350 ml min^{-1}	Mass spectrometer, EI, 70 eV, transfer line = 280 °C, ion source = 230 °C, scan range 50–500 AMU

analyte should be achieved *via* visual inspection of the chromatogram (retention times), mass spectra and comparison with the NIST, WILEY or SWGTOX databases. This is followed by the calculation and comparison of the relative retention factor (retention time of the analyte divided by the retention time of the internal standard). Results should be considered tentative until further analysis of the standard using the same method and a comparison of the results is made. Confirmation of the identity is achieved by comparison of mass spectra, parent and daughter ions and their relative intensities.

Total ion chromatograms (TICs), extracted ion chromatograms (EICs), or selected ion monitoring (SIM) can be used for quantification. A total ion chromatogram is obtained when analysis is conducted in scan mode, *i.e.*, scanning over the required mass range, usually around 45 to 500 AMU. Once the TIC is obtained, if analysts extract a mass ion of interest, this is called an extracted ion chromatogram. Peak areas or heights of either the whole peak, a mass ion extracted from the peak, or a mass ion selectively monitored can be used for quantification, along with equivalent results from an internal standard.

12.2.6 High-performance Liquid Chromatography-based Techniques

HPLC is the method of choice for analysing benzodiazepines. This technique does not expose analytes to high temperatures, which result in thermal decomposition of the benzodiazepines, eliminating the need for derivatisation reactions. A simple sample preparation method such as dissolution followed by filtration or centrifugation is usually sufficient for analysing seized samples. While detection by UV spectroscopy coupled with HPLC (HPLC-UV) remains common, coupling with mass spectrometry (*i.e.* LC–MS) is the preferred method for identification and quantification studies.

Due to the diversity in the chemical structure of benzodiazepines, the use of one HPLC method to separate all benzodiazepines might not be possible. Additionally, there are differences in the types of benzodiazepines seized in different countries, and hence, the HPLC or liquid chromatography-based methods used may differ depending on the circulation of benzodiazepines in the country. An exemplar method that works is one involving a slight modification from the method given in the UNODC manual,[3] which involves a Spherisorb ODS-2 column (25 cm × 4.6 mm i.d. 5 µm particle size) with a mobile phase of methanol/water/phosphate buffer (0.1M) at 55 : 25 : 20 by volume, pH 7.25, delivered at a flow rate of 1.5 ml min^{-1}. Detection is with UV detection at 240 nm. As with GC methods, confirmation of identity can only be made when standards and samples are chromatographed under the same conditions and batch of buffer, and retention times, diode array spectra or mass spectra are compared and shown to be identical.

References

1. R. Abouchedid, T. Gilks, P. I. Dargan, J. R. Archer and D. M. Wood, *J. Med. Toxicol.*, 2018, **14**, 134.
2. A. Grela, L. Gautam and M. D. Cole, *Forensic Sci. Int.*, 2018, **292**, 50.
3. Available from: https://www.unodc.org/documents/scientific/barbiturates_and_benzodiazepines.pdf. Accessed 19 Aug 2024.

13 The Analysis of Fentanyl and Its Analogues

13.1 Introduction

Fentanyl and its analogues are powerful synthetic opioids that are becoming increasingly abused, especially when mixed with opiate drugs such as heroin. Although fentanyl is abused on its own, it is commonly detected as an additive in heroin, methamphetamine and cocaine. The potency of this group of drugs (fentanyl is 50–100 times more potent than morphine and carfentanil 10 000 times more potent than morphine) (Table 13.1), ease of synthesis, the ease with which precursors can be obtained and the ready availability of equipment required for synthesis have led to the growth of the use of fentanyl and analogues in the recreational drug market. In the 1970s and 1980s, drugs containing fentanyl and related compounds appeared in the drug market, and, more recently, clandestinely synthesised fentanyl analogues have reappeared, resulting in many accidental overdose and death cases in North America and Europe. Fentanyl was introduced to the market in the early 1960s, followed by sufentanil 1980, alfentanil in 1984, methylfentanyl analogues in 1988, and hydroxy and thio analogues in 1990, with other analogues being continuously developed[1] (Figure 13.1). These drugs have also been detected in Asia and Africa. Since 2011, there has been a significant rise in both fatal overdoses linked to the improper use of illegally manufactured fentanyl and its analogues in the USA. As with designer benzodiazepines, many fentanyl analogues are synthesised solely for the recreational drug market, but licit fentanyl pharmaceutical products for human and veterinary use are also diverted *via* theft and fraudulent prescriptions to be misused

Table 13.1 Relative potency of fentanyl analogues as compared to the potency of morphine as 1. Data from ref. 1.

Compound	Relative potency	Analgesic activity, ED_{50}/ mg kg^{-1}
Furanylfentanyl	7	0.02
Parafluorofentanyl	16	0.021
alpha-Methylfentanyl	56.9	0.0058
Alfentanil	75	0.044
Acrylfentanyl	170	0.082
Remifentanil	220	0.73
Fentanyl	224	0.062
Sufentanil	4520	0.00071
(+) *cis*-3-Methylfentanyl	5530	0.00058
Carfentanil	10 030	0.00032

in the illegal drug market. While many are used solely as recreational drugs, sufentanil, alfentanil and remifentanil have been approved as pharmaceutical compounds.[1]

Fentanyl consists of three primary components: a phenylalkyl group, a piperidinyl ring, and a propylalkylamide group (Figure 13.2). Analogues of fentanyl are formed by modifications to the propylalkylamide moiety, adjustments to the chain length in the arylalkylpiperidinyl group, or substitutions of the aromatic substituent in the phenylalkyl moiety. A number of fentanyl derivatives have been developed by the legitimate pharmaceutical industry by adding different substituents into the core structure molecule to enhance potency. However, clandestine chemists have also mimicked this strategy to synthesise illicit derivatives.

Some of these compounds may exist as enantiomers, and their potency may differ. An example would be the isomers of 3-methylfentanyl that exhibit different analgesic potencies. The *trans* isomer showed slightly higher activity than fentanyl, while the *cis* form showed eightfold increased potency.

Fentanyl was placed under international control as a Schedule I substance in 1964 under the Single Convention on Narcotic Drugs 1961. Since then, all fentanyl pharmaceutical analogues approved for medical use (*e.g.* sufentanil, alfentanil, and remifentanil) and the analogues that have not been developed into pharmaceutical products have been controlled at different times.[1] In the United Kingdom, fentanyl and its analogues are classified as Class A controlled substances through the Misuse of Drugs Act 1971. Fentanyl is a Schedule II narcotic under the United States Controlled Substances Act 1970, the Controlled Substances Analogue

The Analysis of Fentanyl and Its Analogues 195

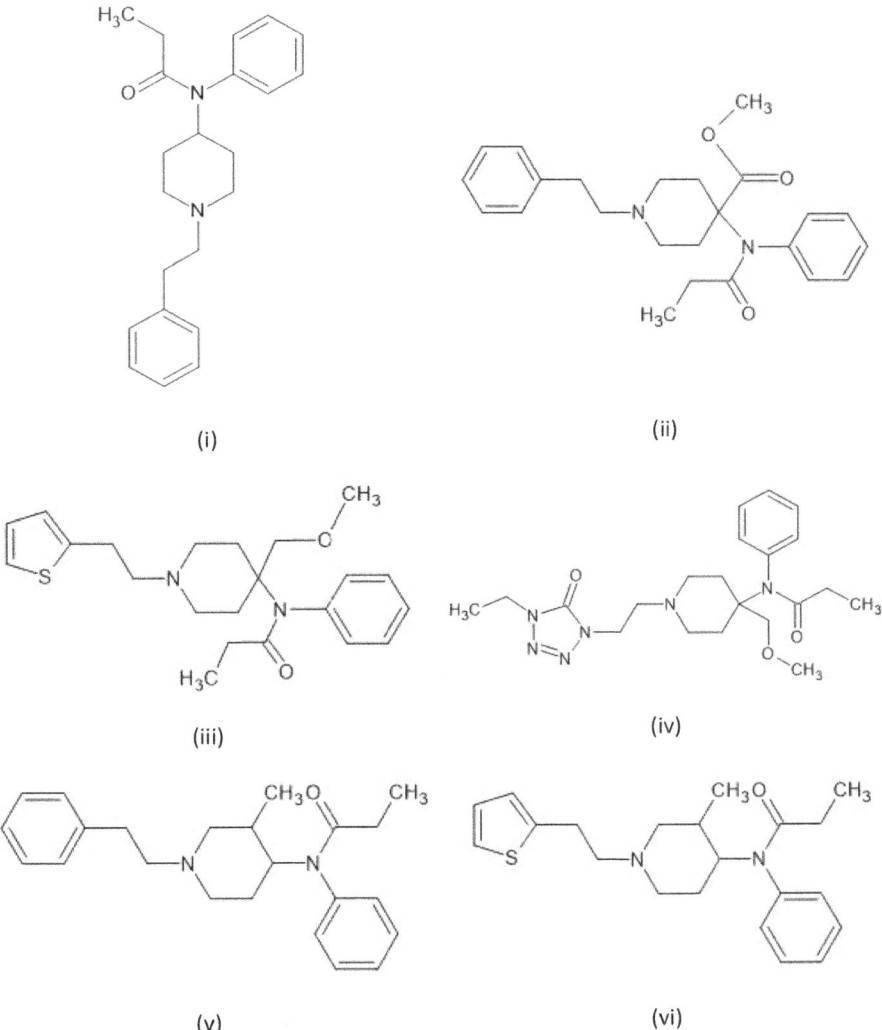

Figure 13.1 Examples of the early fentanyl compounds found in the illicit drug market: (i) fentanyl, (ii) carfentanil, (iii) sufentanil, (iv) alfentanil, (v) 3-methylfentanyl, and (vi) 3-methylthiofentanyl.

Enforcement Act 1986 (Federal Analogue Act) and state laws. An emergency class-wide scheduling of fentanyl-related substances, based on the core structure, was introduced in 2018 in the USA. In 2023, the DEA controlled fentanyl-related substances in Schedule I of the CSA as a chemical structural class.

Fentanyl analogues including fentanyl-mixed heroin, have various street names, such as Apache, Chiclets, China Girl, China Town,

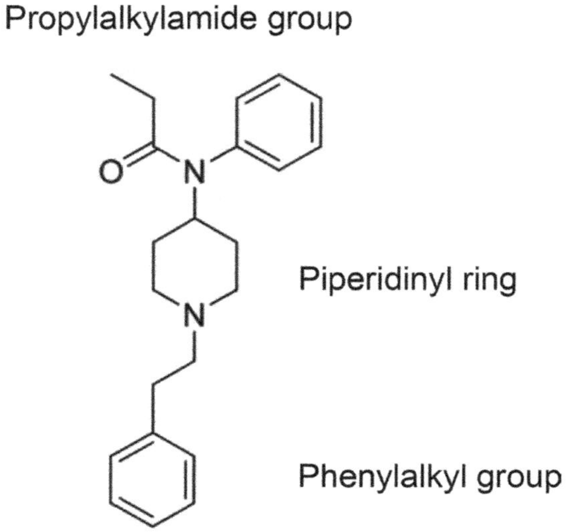

Figure 13.2 Component groups of the core fentanyl molecule.

Dance Fever, Friend, Goodfellas, Great Bear, He-Man, Jackpot, King Ivory, Murder 8, Perc-O-Pops, Tango & Cash, Redrum (murder spelled backwards), White Heroin, White China, and TNT. Fentanyl abuse, including the use of drug mixtures where fentanyl is mixed with other substances, occurs *via* diverse methods such as injection, inhalation, ingestion, and dermal absorption. Additionally, pharmaceutical fentanyl patches are misused by extracting their contents and then administering them. Illicitly manufactured fentanyl and carfentanil are encountered in powder form or as counterfeit tablets. Due to the potency of these drugs, even slight errors in dosage can be fatal, and in general, there is an increased possibility of overdose arising from heroin cut with these synthetic opioids. As an example, the majority of overdose fatalities associated with opioids in North America have been attributed to illegally manufactured fentanyl.

13.2 Analytical Methods for the Analysis of Fentanyl-type Drugs

13.2.1 Sampling of Fentanyl-type Drugs

Due to the potency of these drugs and the very small amounts that can cause lethal overdoses, ensuring the safety of officers and researchers becomes a very important first step and cannot be overemphasised. It is very easy to contaminate a laboratory surface with powdered drug samples,[2] and this should be avoided, especially where fentanyl and related drugs are analysed. A full risk assessment should be undertaken before handling these drugs. Second, there is a pressing need for rapid, cost-effective, *in situ* testing methods that allow for the detection and differentiation (due to differing potency) of fentanyl analogues. The UNODC-recommended methods published in 2017[1] cover the methods for the identification of fentanyl and its analogues in biological samples. However, at the time of writing this chapter, the authors were unable to find similar guidelines for the analysis of seized samples and so the general principles of drug analysis need to be applied. Most authors of the papers that describe suitable analytical methods have reported the use of a polythetic approach (*i.e.* where no single method will prove the identity of the drug, but where collectively the use of a number of methods points to one drug only) to increase the discrimination power and identification of these drugs. Once the health and safety of analysts has been ensured, the sampling protocols described in Chapter 2 suffice for the identification of samples to be analysed.

13.2.2 Presumptive Testing of Fentanyl-type Drugs

A colour test specific for fentanyl and acetyl-fentanyl detection (NARK Fentanyl Reagent) has been described, which turns orange/red in the presence of fentanyl and blue in the presence of heroin.[3] This leads to the possibility of an ambiguous colour reaction when heroin is mixed with fentanyl or acetyl-fentanyl. The Fen-Her Fentanyl/Heroin Test Kit states that it gives a positive reaction with fentanyl freebase, carfentanil, acetyl fentanyl, fentanyl citrate, methoxyacetyl fentanyl, furanyl fentanyl, α-methylfentanyl, *trans*-3-methyl fentanyl and *cis*-3-methyl fentanyl. Fentanyl produces a positive reaction with aqueous cobalt thiocyanate (200 mg cobalt(II) thiocyanate in 10 ml water) and Young's test [aqueous cobalt thiocyanate with an additional step of adding one drop of $SnCl_2$ reagent (500 mg tin(II) chloride in 10% v/v HCl in water)]. In both cases, the material or solution turns blue.[4] However, with cobalt isothiocyanate (Section 13.3.2) reagent, after adding hydrochloric acid, samples containing fentanyl will turn blue and remain blue.

13.2.3 Instrumental Methods for Screening for Fentanyl-type Drugs

Ion mobility spectrometry (IMS) utilises the speed of ions moving through a carrier gas to separate and detect them. The mobility of ions relies on their charge, reduced mass, and collisions. Commercial off-the-shelf (COTS) ion mobility spectrometers (IMS) have been used in the field (*in situ*) to screen for illegal substances and recreational drugs. However, volatility issues related to the salt and free base forms of fentanyl and its analogues prevent differentiation between these forms or between drugs with different salt forms using IMS.

This is a difficult and emerging group of drugs, and researchers have also used a combination of analytical techniques such as FTIR, Raman and testing kits[5] to screen street samples for the presence of fentanyl and related drugs. This approach enhances the ability to identify and differentiate fentanyl-containing drugs. Using infrared spectroscopy, it has been possible to achieve detection where fentanyl-type drugs represent 3–4% of the weight of the drug, with a relatively high specificity (90.2%). Handheld Raman spectrometers offer ease of use with "point and shoot" analysis, real-time database searches, and a detection limit reported at 25 µg ml^{-1} for fentanyl and its analogues.[5] A semi-quantitative method utilising surface-enhanced Raman scattering (SERS) with a handheld Raman instrument has been employed.[6] While fentanyl identification was successful, calibration proved challenging in the presence of diluents. Differentiation between fentanyl and various fentanyl analogues (butyrylfentanyl, furanylfentanyl, acetylfentanyl, and ocfentanyl) (Figure 13.3) was also difficult, indicating the limitations of the technique.

A portable GC–MS has been used to analyse seized drugs[7] and was found to be suitable for screening analysis. The authors analysed 24 drugs, including fentanyl, using the FLIR Griffin™ G510 quadrupole GC–MS. The method was validated and seems promising, with an accuracy of close to or at 100%.

It is clear that there are a number of methods available for screening drug samples for the presence of fentanyl and its analogues. Of paramount importance is the safety of the personnel involved. On this basis, the choice of method depends, as articulated in the preface to this book, on the resources available and the training of the personnel involved in the screening process. On this

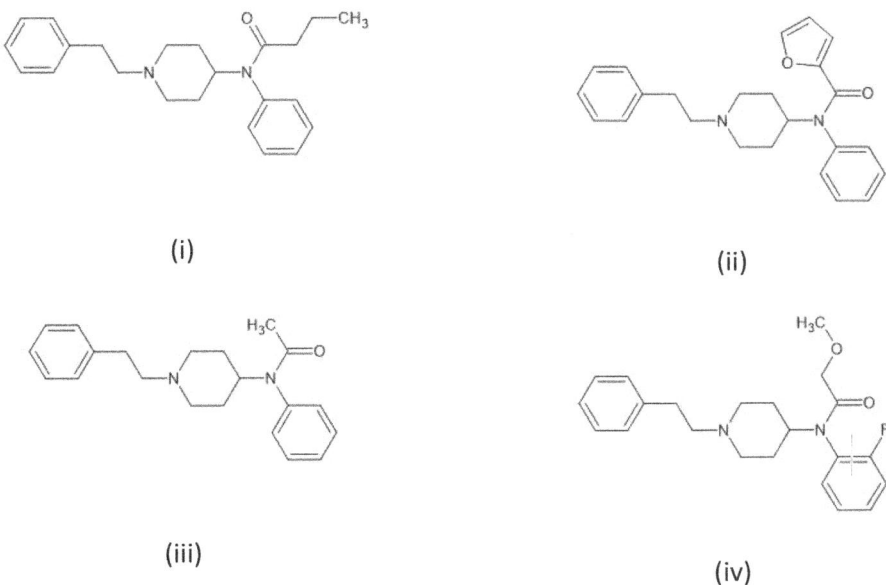

Figure 13.3 Structure of (i) butyrylfentanyl, (ii) furanylfentanyl, (iii) acetylfentanyl and (iv) ocfentanyl.

basis, readers are recommended to consult the literature and make appropriate choices.

13.2.4 Confirmation of Identity and Quantification of Fentanyl-type Drugs

For the confirmation of the identity and quantification of fentanyl and its analogues, gas chromatography–mass spectrometry (GC–MS), liquid chromatography–mass spectrometry (LC–MS), tandem mass spectrometry (MS/MS), and high-resolution mass spectrometry (HR-MS) have been employed. A LC-QTOF (Table 13.2) with a dilute-and-shoot method, along with GC–MS (Table 13.3) to analyse seized drugs, including fentanyl analogues and adulterants, has been used.[7] Samples were dissolved in methanol or ethanol and analysed using GC–MS. For LC-QTOF, diluted samples prepared in the mobile phase (95% aqueous and 5% organic) were analysed. It is noteworthy that analytical methods such as GC–MS, LC–MS/MS, HR-MS, and Q-TOF recommended by the UNODC[1] for the analysis of fentanyl analogues in biological samples could equally be applied to seized drugs suspected to contain fentanyl-type drugs. The only difference would be that the levels of drugs expected in street samples might be different to those contained in biological matrices.

Results are obtained as chromatographic data and mass spectra. Therefore, in terms of data interpretation, identification should include the use of retention time, retention index, mass ions, and their relative intensities. Table 13.4 provides exemplar characteristic mass ions for commonly encountered pharmaceutical fentanyl analogues.

In conclusion, fentanyl and its analogues represent a significant challenge in the field of illicit drug use due to their potency and involvement in overdose deaths. There is currently no recommended methodology for street samples suspected to contain fentanyl, but the application of the principles of drug analysis for court-going purposes should suffice. Various analytical techniques, ranging from traditional methods including colour tests through to advanced forms of chromatography and tandem mass spectrometry, can be used. Furthermore, the emergence of novel analogues continues to pose challenges for law enforcement and public health agencies worldwide, and there is a need for further analytical methodologies to be explored, especially given the health and safety risks that this group of compounds presents.

Table 13.2 QTOF (Triple TOF 5600+ mass spectrometer from Sciex coupled with a Shimadzu LC-QTOF, positive ESI) parameters for the study of seized drugs including fentanyl.[7]

Parameter	Setting
Mode	Reverse-phase gradient elution
Column	Phenomenex Kinetex C18 analytical column (50 mm × 3.0 mm, 2.6 mm) maintained at 30 °C
Mobile phase	Ammonium formate (10 mM, pH 3) [A]; 0.1% formic acid in methanol–acetonitrile (50 : 50) [B]
Gradient	70A : 30B, transitioned to 5A : 95B over 1.5 min and held until 2.3 min, before returning to 70A : 30B at 2.4 min and held until the final run time of 3 min
Flow rate	0.5 ml min^{-1}
Scan	Precursor ions were acquired by positive ESI scan (100–510 Da)
Collision energy	Fragmentation of precursor ions was achieved using 35 +/− 15 eV

Table 13.3 GC–MS parameters for seized drugs including fentanyl analogues.[7]

Parameter	Setting
Instrument	Agilent GC–MS, EI mode
Injection port/°C	265
Column dimension	J&W DB-1 capillary column (12 m × 0.20 mm × 0.33 mm)
Oven temperature	50 °C to 340 °C at 30 °C min^{-1}, hold for 2.33 min
Flow rate	1.2 ml min^{-1}
Injection	Splitless, 1 µl
Transfer line/°C	300
Scan range	m/z = 40–500

Table 13.4 Characteristic mass spectral ions (GC–MS) for selected fentanyl-based compounds.[1]

Drug	Characteristic mass ions (m/z)
Fentanyl D-5	250 and 194
Fentanyl	245, 146, 189, and 202
Sufentanil	140, 187, 238, and 289
Alfentanil	289, 222, 268, and 359

References

1. Available from: https://www.unodc.org/unodc/en/scientists/recommended-methods-for-the-identification-and-analysis-of-fentanyl-and-its-analogues-in-biological-specimens.html. Accessed 19 Aug 2024.
2. E. Sisco, M. E. Staymates and L. M. Watt, *Forensic Chem.*, 2020, **20**, 100259.
3. Available from: https://www.sirchie.com/nark-ii-fentanyl-reagent.html. Accessed 15 Aug 2024.
4. P. A. Kosecki, P. Brooke and E. Canonico, *J. Forensic Sci.*, 2022, **67**(5), 2082.
5. T. C. Green, J. N. Park, M. Gilbert, M. McKenzie, E. Struth, R. Lucas, W. Clarke and S. G. Sherman, *Int. J. Drug Policy*, 2020, **77**, 102661.
6. M. Smith, M. Logan, M. Bazely, J. Blanchfield, R. Stokes, A. Blanco and R. McGee, *J. Forensic Sci.*, 2021, **66**(2), 505.
7. T. R. Fiorentin, A. J. Krotulski, D. M. Martin, T. Browne, J. Triplett, T. Conti and B. K. Logan, *J. Forensic Sci.*, 2019, **64**(3), 888.

14 The Analysis of Phencyclidine and Ketamine

14.1 Introduction

Phencyclidine (1-(1-phenylcyclohexyl)piperidine, PCP) (Figure 14.1) is a hallucinogenic and dissociative substance which used to be administered for its anaesthetic properties. PCP can induce a trance-like state, sedation, memory loss, and provides pain relief. It is sold in various forms, including powder, crystals, tablets, capsules and liquid (solution). In its oil form, it is yellow; however, crystals and powder range in colour from white to light brown. PCP has been identified as an adulterant in seizures of cocaine and heroin, but it is also found in street samples as a single drug.

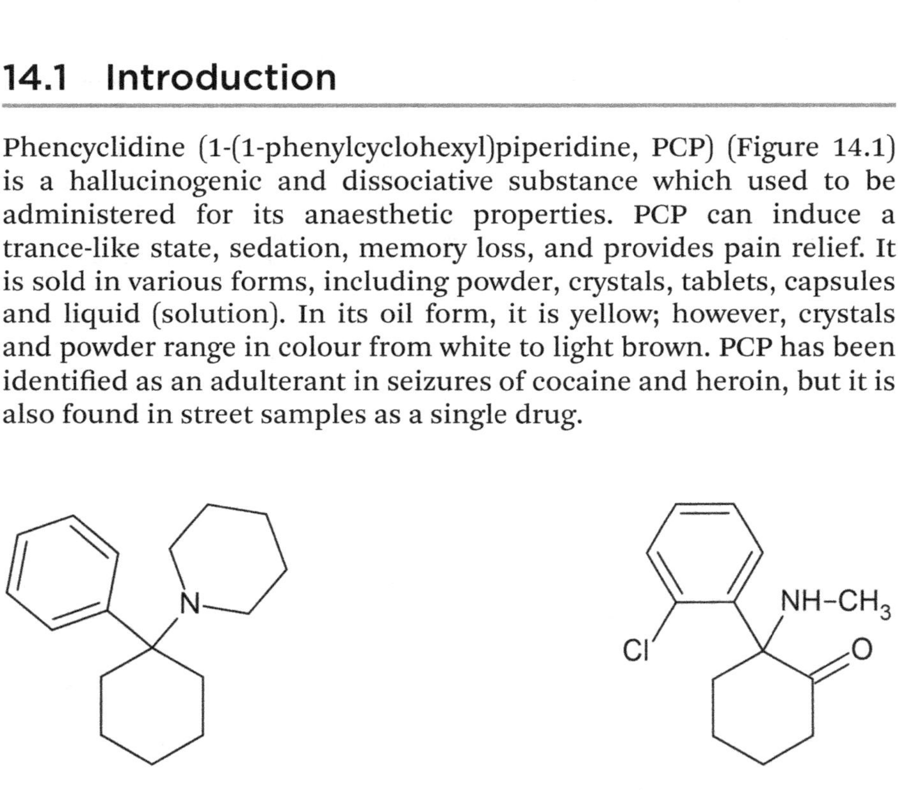

Figure 14.1 Structure of (i) phencyclidine (PCP) and (ii) ketamine, illustrating their structural similarity.

Controlled Drug Analysis
By Michael D. Cole, Lata Gautam and Agatha Grela
© Michael D. Cole, Lata Gautam and Agatha Grela 2025
Published by the Royal Society of Chemistry, www.rsc.org

Ketamine (2-(2-chlorophenyl)-2-(methylamino)cyclohexan-1-one) (Figure 14.1) is an anaesthetic drug with dissociative and hallucinogenic properties. It was first introduced in the 1960s as a substitute for phencyclidine. Ketamine can cause numbness, a feeling of detachment from the body, memory loss, feelings of happiness and relaxation, as well as visual and auditory hallucinations. For clinical

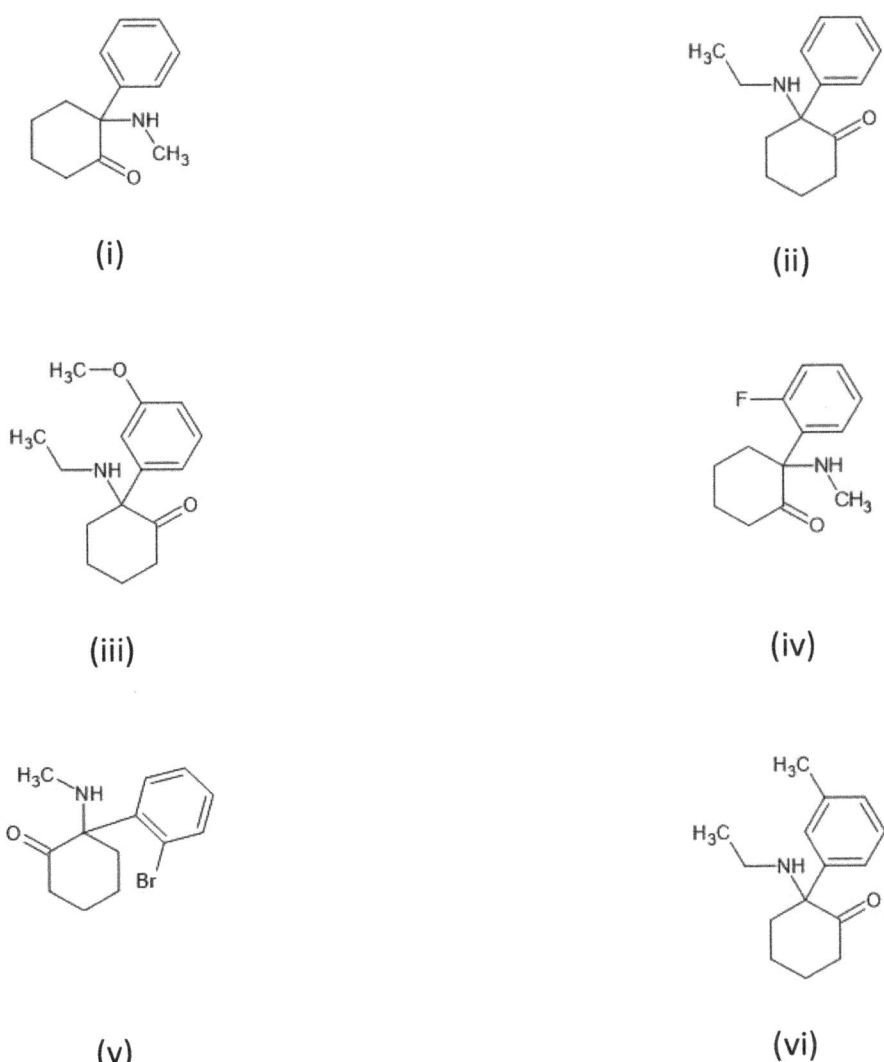

Figure 14.2 Structure of (i) deschloroketamine, (ii) deschloro-*N*-ethylketamine, (iii) methoxetamine (MXE), (iv) fluoroketamine, (v) bromoketamine and (vi) deoxymethoxetamine.

purposes, ketamine is usually a clear liquid. In the recreational drug market, illicit ketamine is sold in crystalline powder with colours ranging from white to brown. There are currently over 20 ketamine analogues identified, which have a 2-phenyl-2-aminocyclohexone structure as shown in the often encountered exemplar compounds (Figure 14.2).

In the UK, under the Misuse of Drugs Act 1971, ketamine is controlled as a Class B drug, while phencyclidine is listed as a Class A drug. In the USA, under the Controlled Substances Act, ketamine is a Schedule III non-narcotic substance, whereas phencyclidine is a Schedule II non-narcotic substance. In 2015, the decision to include ketamine in Schedule IV of the UN 1971 Convention was postponed – it is currently unscheduled. Phencyclidine is controlled as a Schedule II drug under the UN 1971 Convention.

14.2 Analysis of Phencyclidine and Ketamine

14.2.1 Sampling and Presumptive Tests for Ketamine and PCP

Whatever the number and dose form of these drugs, a representative sample must be obtained for analysis. The protocols in Chapter 2 provide guidance on how such samples may be obtained.

14.2.2 Presumptive Tests for Phencyclidine and Ketamine

One of the colour reagents used to detect ketamine is Zimmermann's reagent [a 1% solution of 1,3-nitrobenzene in methanol (w/v; 2 drops), which is then basified by the addition of 2 drops of a second solution of 15% potassium hydroxide in water (w/v)]. Ketamine gives a light pink to purple colour reaction. It should be noted that a number of drug classes react with Zimmermann reagent and give similar colour reactions.

PCP does not react with Zimmermann reagent. However, phencyclidine hydrochloride will produce a strong greenish-blue colour with a 1% aqueous solution of cobalt(II) thiocyanate. PCP will also produce a pink colour when tested with Mecke's reagent. As a range of drug classes react with these reagents, it is necessary to compare the colour reactions obtained with those from standard compounds. Negative controls should also be undertaken to demonstrate the cleanliness of the reagents and equipment.

14.2.3 Microcrystal Tests for Phencyclidine and Ketamine

Microcrystal tests can also be used to screen for ketamine.[1] Using a 5% platinic iodide solution to test the sample that has first been dissolved in water will result in the formation of rhomboidal plates. Similarly, microcrystal tests can be used to screen for PCP.[2] For PCP, two reagents can be used: (i) acidified potassium permanganate (2% KMnO4 in water, w/v, with 5 drops of concentrated H_3PO_4 with the sample dissolved in 20% acetic acid), which results in crystals with a number of forms, and (ii) ammonium thiocyanate, which also results in crystals of a variety of forms.

14.2.4 Thin Layer Chromatography of PCP and Ketamine

A number of simple TLC systems can be used to separate these drugs from recreational drug mixtures. Examples are given in Table 14.1. Ketamine requires that the stationary phase is activated (dried out) to remove any sorbed water prior to analysis and is basified with potassium hydroxide. Basifying the stationary phase ensures that the drug, if present as a salt, will be converted into the free base form, improving chromatographic properties and reducing tailing. Following the development of the chromatograms, they should be visualised under UV light (254 nm) and then with acidified iodoplatinate spray or Dragendorff reagents. It should be noted that these

Table 14.1 TLC stationary and mobile phases for the analysis of PCP and ketamine.

	Ketamine	PCP
Stationary phase	Activated silica gel G plates 250 µm thick (washed with 0.1 M potassium hydroxide in methanol)	Silica gel GF plates 250 µm
Mobile phase(s)	Methanol : concentrated ammonia (100 : 1.5)	Chloroform:acetone (9 : 1)
	Chloroform : methanol (9 : 1)	Chloroform-saturated methanol with ammonia (18 : 1) (visualisation also with Dragendorff reagent)
	—	Chloroform:methanol (9 : 1) (visualisation also with Dragendorff reagent)

reagents react with a wide range of alkaloids, and as a consequence, TLC can be used to eliminate the presence of certain ketamines or PCP, but not to prove the identity of any suspected controlled substance. Definitive identification of PCP or ketamine requires the use of instrumental methods.

14.2.5 Instrumental Analysis of Phencyclidine and Ketamine

14.2.5.1 Gas Chromatographic Methods

Suspected or known samples of powdered ketamine can be analysed by GC-FID or GC-MS (Table 14.2) following dissolution of 10 mg of the sample in 2 ml of 0.5% ethanolic solution of an internal standard, *e.g.* dieldrin.[3] Any insoluble material should be removed by centrifugation prior to analysis. Tentative identification is made on the basis of retention time, while definitive identification can be made using GC-MS using the mass spectrum and comparison of the data from the sample with those of a standard.

If ketamine analogues are suspected to be included in the seized samples, the following method (Table 14.3) can conveniently be used.

14.2.5.2 Spectroscopic Methods for the Forensic Analysis of PCP and Ketamine

New technologies are emerging that aim to facilitate the on-the-spot/*in situ* analysis of suspected illicit drugs. One of them is the

Table 14.2 Operating conditions for the analysis of ketamine by GC-FID or GC-MS.

Parameter	GC-FID method			GC-MS method		
Column	HP-5 (30 m × 0.32 mm, 0.25 µm)			HP-5MS (30 m × 0.25 mm, 0.25 µm)		
Carrier gas	Hydrogen			Helium		
Flow rate	1 ml min^{-1}			1 ml min^{-1}		
Injector port	Data not available; split according to concentration			300 °C, splitless		
Injection volume	1 µl			1 µl		
Oven programme	Ramp (°C min^{-1})	Temperature (°C)	Hold (min)	Ramp (°C min^{-1})	Temperature (°C)	Hold (min)
	—	210	1	—	100	2.25
	20	270	0	40	180	0
	20	300	1	10	300	10
Detector	FID, settings not available			m/z range: 50–550 transfer line: 300 °C		

Table 14.3 Operating parameters for the GC–MS analysis of suspected ketamine analogues.

Parameter	Setting		
Column	HP-5MS (30 m × 0.25 mm, 0.25 μm)		
Carrier gas	Helium		
Flow rate	1 ml min^{-1}		
Injector port	275 °C, split 10 : 1		
Injection volume	1 μl		
Oven programme	Ramp (°C min^{-1})	Temperature (°C)	Hold (min)
	—	60	0
	20	280	0
Detector	*m/z* range: 50–550		
	Transfer line: 300 °C		
	Ion source: 150°C		
	Quadrupole: 230 °C		

application of near-infrared analysers, which allow for the quick identification of drugs contained in powder form, including ketamine, where the spectrum is often obtained through the packaging of the drug found at the scene. This has the added advantage that there is a reduced risk of sample contamination and to the health and safety of the analyst on site. This technology has already been successfully applied in forensic casework.[4] This analytical platform compares spectra in the library with the spectrum collected from the sample and can provide information on whether the drug contains a pure drug compound, a single drug compound in a mixture, a mixture of drugs or whether the analysis was inconclusive. The same technology would be applicable for the detection of PCP, provided that reference spectra are available and uploaded onto the platform. The spectrum will differ depending on the purity of the sample and also the form of the drug: free base or salt. Drug identification can be made provided that an appropriate spectral library is available.

Raman spectroscopy has been successfully used in the detection of ketamine.[5] Such identifications can be made using laboratory benchtop or handheld devices. In 2017, the UNODC released guidelines on the use of handheld Raman devices for the analysis of seized material.[6] As with infrared spectroscopy, an appropriate library is required, ideally obtained on the same spectrometer, because of the difference in the Raman spectrum obtained on different instruments due to the differences in Raman spectrum acquisition due to differences in the laser type and strength.

14.2.6 Identification of New Analogues of Ketamine

As with synthetic cannabinoid receptor agonists, ketamine and its analogues can be analysed using NMR,[7] which will provide structural information about the compound. This will be particularly useful for the identification of new analogues and impurities. For that, the purified sample needs to be dissolved in deuterated trichloromethane ($CDCl_3$), filtered or centrifuged to remove particulate material and then submitted for analysis.

References

1. M. P. Elie and L. E. Elie, Microcrystalline tests in forensic drug analysis, *Encycl. Anal. Chem.*, 2009, **1**, 471.
2. K. M. Brinsko, D. Golemis, M. B. King, G. J. Laughlin and S. B. Sparenga, in *A modern compendium of microcrystal tests for illicit drugs and diverted pharmaceuticals*, McCrone Research Institute, Chicago, 2015.
3. E. Bertol, L. Bigagli, S. D'Errico, F. Mari, D. Palumbo, J. P. Pascali and F. Vaiano, *Forensic Sci. Int.*, 2018, **284**, 194.
4. R. F. Kranenburg, H. J. Ramaker and A. C. van Asten, *Forensic Sci. Int.*, 2022, **341**, 111467.
5. M. J. West and M. J. Went, *Drug Test. Anal.*, 2011, **3**(9), 532.
6. Available from: https://www.unodc.org/documents/scientific/SCITECT-26_Addendum_2021_Eng.pdf. Accessed 16 Aug 2024.
7. G. Frison, L. Zamengo, F. Zancanaro, F. Tisato and P. Traldi, *Rapid Commun. Mass Spectrom.*, 2016, **30**(1), 151.

Subject Index

1-acetyl-lysergic acid diethylamide test 72
1-benzyl-4-methylpiperazine 156
1-benzylpiperazine 156
1-phenylpiperazine 156
1-propionyl-lysergic acid diethylamide test 72
1,2-naphthoquinone-4-sulfonic acid sodium salt (NQS) reagent 160–161
1-(3-chlorophenyl)-4-(3-chloropropyl)piperazine 156
1-(3-chlorophenyl)piperazine 156
1-(3-trifluoromethylphenyl)piperazine 156
1,4-benzodiazepine 183
1-(4-bromo-2,5-dimethoxybenzyl)piperazine 156
1-(4-chlorophenyl)piperazine 156
1-(4-fluorophenyl)piperazine 156, 158
1-(4-methoxyphenyl)piperazine 156
2,5-dimethyl-3,6-diphenylpyrazine 94
3,4-methylenedioxymethamphetamine (MDMA) 125–127, 133, 136, 138, 143, 170
   diode array spectra 136
   inorganic ions analysis 142–143
3,4-methylenedioxyamphetamine (MDA) 125–127, 133, 136
3,4-methylenedioxyethyl amphetamine (MDEA) 126, 127, 133, 136, 138
3,4-methylenedioxypyrovalerone (MDPV) 88, 89, 92, 93, 95, 98, 99

3-chloromethcathinone (3-CMC) 89
3-methylfentanyl 195
3-methylthiofentanyl 195
4-benzylpyrimidine 129
4-chloromethcathinone (4-CMC) 89
4-methyl-5-phenylpyrimidine 129
4-methylphenylpiperazine 156
6-monoacetylmorphine (6-MAM), chromatogram 116

Δ^9-tetrahydrocannabinol (Δ^9-THC) 27–28, 40, 48–49
Δ^9-tetrahydrocannabinolic acid (THCA) 28, 43

AB-001 50, 51
acetaminophen, chromatogram 116
acetylcodeine 113, 115, 116
acetylfentanyl
   colour test 197
   structure 199
alfentanil 193–195, 201
alkaline beam test, cannabis 38
allobarbital 149, 150
AM-2201 58, 59
amobarbital 149, 150
amphetamine-type drugs 88
   adulterants 127
   chemical impurity profiling 143–146
   description and sampling 131–132
   diluents 127
   diode array spectra 136
   doses 127
   gas chromatography 138–142

high-performance liquid
 chromatography 135–138
inorganic ions analysis 142–143
isomers separation 136–137
Leuckart synthesis 128
nitrostyrene route 130–131
powdered samples 131
powder form 126
presumptive tests 132–133
reductive amination 130
ring-substituted variants 125
structure 126
synthesis 127–131
tabletted dose forms 131–132
thin layer chromatography
 133–135
amplified fragment length
 polymorphisms (AFLP) 179
arbitrary sampling regimes 21
aziridine 131

barbital 149, 150
barbiturates
 5,5 disubstituted 148–149
 barbituric acid 148–149
 forensic science 148
 gas chromatography 152–154
 high-performance liquid
  chromatography 151–152
 injectable forms 149–150
 legislative classification 150
 NMR spectroscopy 154
 powdered and tablet drugs 150
 presumptive tests 150
 thin layer chromatography 151
barbituric acid 148–149
bentazepam 182, 183
benzodiazepines 4, 105, 116, 148,
 193
 categories 181–184, 186
 designer 181–182
 dissolution 187–188
 gas chromatography-based
  analysis 189–191
 high performance liquid
  chromatography 191–192
 physical description 186–187

presumptive tests 187
sampling 186–187
thin layer chromatography
 188–189
traditional vs. designer 185
benzoyl ecgonine 169, 175
benzoylindoles 51
benzoylmethylketoxime 130
benzylpiperazine (BZP) 155–156
bis(trimethylsilyl)trifluoroacetamide
 (BSTFA) 76–77, 176
bromoketamine 203
brotizolam 183
bulk sample treatment 16–17
butabarbital 149
butyrylfentanyl 199

caffeine 109, 116, 127, 157, 166, 169,
 182
Calvin cycle 45
cannabinol 15, 27–28, 37, 39, 41
cannabis and products 13
 active drug 27–28
 bulk vs. trace samples 30
 cannabinol 28–29
 chemical analysis 36–47
 edibles 29–30
 hash oil 34
 herbal material 31–32
 photo-oxidation 33
 physical description 30–35
 presumptive tests 36–38
 resin 30–31
 resinous material 32–34
 sample weight 34–35
 sampling 35–36
 thin layer chromatography 38–40
carfentanil 193–197
cathine 88, 92, 93
cathinones
 and amphetamines, similarities
  89
 analogues 88–89
 decomposition 94
 enantiomers 91
 legislative control 91–92
 profiling 101–102

structure 88
synthetic (*see* synthetic cathinones)
tautomerism 91
chloroform 13, 18, 36, 37, 40, 68, 78, 93, 109, 110, 114, 139, 162, 172–174, 176
clonazolam 184, 185
clotiazepam 184
cobalt thiocyanate test, cocaine 172
cocaine 1, 13, 15, 92, 105, 202
    classification 168
    cocaine hydrochloride 168
    colour tests 172
    crack cocaine 168
    DNA analysis 179
    extraction and synthesis 170
    gas chromatography–mass spectrometry 175–176
    high-performance liquid chromatography 176–177
    infrared spectroscopy 178
    microcrystalline tests 172–173
    presumptive tests 172
    sample preparation 173–174
    sampling 171–172
    speedballing 169
    stable isotope analysis 179–180
    thin layer chromatography 174–175
    ultraviolet spectroscopy 177–178
Controlled Substances Act (CSA) 6–7, 194, 204
Controlled Substances Analogue Enforcement Act 1986 194–195
Corinth IV salt test, cannabis 36, 37
CP47,497 49, 50
crack cocaine 168, 170
cyclic voltammetry, piperazines 166
cyclobarbital 149, 150
cyclohexylphenols 48, 49

deoxymethoxetamine 203
deschloroketamine 203
deschloro-*N*-ethylketamine 203
designer benzodiazepines 181–182, 185, 187, 188, 193

diacetylmorphine, chromatogram 116
diamorphine 2, 15, 104–107, 111–113, 119, 123
    DNA analysis 123
    GC–MS analysis 117
    NMR spectroscopy 118
diazepam 105, 157, 181–183, 185, 189
dibenzylpiperazine (DBZP) 157, 165
diclazepam 181, 182
Dille–Koppanyi test, barbiturates 150
diluents 103, 109, 127, 142, 166
p-dimethylaminobenzaldehyde (DMAB) 67
dissolution 15, 25, 96, 160
    benzodiazepine 187–188
    breakdown products of drugs 15
    solvents 15
DMSO 72, 117–118
DNA analysis
    cannabis and products 46–47
    cocaine 179
    heroin 123
    psilocybin analysis 78–79
drug control
    in Australia 8
    in India 8
    in United Kingdom 3–6
drug naming system 2–3
drug sampling
    arbitrary sampling regimes 21
    dissolution 25
    drug materials packaging 18–20
    guiding principles 20
    liquid 23–24
    physical description 11–12
    powders 22–23
    statistical sampling 21–22
    tabletted dose forms 24–25
    trace and bulk drug samples 13–18
    wrapping materials 17–18
Drug Trafficking Offences Act 1986 4
Duquenois–Levine test, cannabis 36–37

ecgonine 169, 175, 176
ecgonine methyl ester 169, 173, 175
Ehrlich reagent
    lysergic acid diethylamide test 67
    N,N-dimethyltryptamine 80
    psilocybin analysis 75
eicosane, chromatogram 116
electrochemical methods,
    piperazines 166
ergot alkaloids 68. See also lysergic
    acid diethylamide
etizolam 182, 183, 185

Fen-Her Fentanyl/Heroin Test Kit
    197
fentanyl-type drugs 105
    components 194, 196
    enantiomers 194
    fentanyl-mixed heroin 195–196
    GC–MS analysis 198, 201
    identity and quantification
        199–200
    ion mobility spectrometry 198
    presumptive testing 197
    relative potency 194
    sampling 196–197
    surface-enhanced Raman
        scattering 198
flualprazolam 182, 184
flubromazepam 181, 182, 185
flunitrazepam 181–183, 186, 187
fluoroketamine 203
forensic examination 10
    cathinones (see cathinones)
    piperazines 159–166
Fourier transform infrared
    spectroscopy, synthetic
    cathinones 101
furanylfentanyl 194, 198, 199
furfural test, cannabis 36–38

gallic acid test, amphetamine-type
    drugs 133
gas chromatography (GC)
    barbiturates 152–154
    benzodiazepines 189–191
    mescaline analysis 85

gas chromatography–infrared
    detection (GC–IRD), piperazines
    165–166
gas chromatography–mass
    spectrometry (GC–MS)
    amphetamine-type drugs 138–142
    barbiturates 152–154
    benzodiazepines 189–191
    cannabis and products 42–44
    cocaine 175–176
    fentanyl-type drugs 199–201
    heroin 113–116
    ketamine 206–207
    lysergic acid diethylamide
        69–70
    N,N-dimethyltryptamine 80–81
    piperazines 163–165
    psilocybin analysis 76–77
    synthetic cannabinoids 56–58
    synthetic cathinones 96–98
gas chromatography–flame
    ionization detector (GC–FID)
    benzodiazepines 189–191
    ketamine 206
glandular trichome 31–33

hallucinogen analysis, plants and
    fungi
    dose forms 62
    legislative status 64
    lysergic acid diethylamide
        63–72
    mescaline 82–85
    N,N-dimethyltryptamine 79–82
    psilocybin and psilocin 72–79
hash oil 34, 36
heptafluorobutyric anhydride
    (HFBA) 115, 141–142
herbal cannabis material 24, 31–33
heroin samples 2, 182
    with cocaine 105
    derivatisation 115
    diamorphine 105–106
    DNA analysis 123
    with fentanyl 105
    gas chromatography–mass
        spectrometry 113–116

Subject Index

high-performance liquid
  chromatography 111–113
insufflation 103
intake process 103
mainlining 103
manufacture 106–108
metal ions 120
NMR spectroscopy 116–118
occluded solvents 120–122
origins 106
physical description 108–109
presumptive tests 109
products 104
stable isotopes 118–120
thin layer chromatography 109–111
with tobacco 105
high-performance liquid
  chromatography (HPLC)
  amphetamine-type drugs 135–138
  barbiturates 151–152
  benzodiazepines 191–192
  cannabis and products 40–41
  cocaine 176–177
  heroin 111–113
  lysergic acid diethylamide (LSD)
    68–69
  N,N-dimethyltryptamine (DMT)
    81–82
  piperazines 163
  psilocybin analysis 77–78
  synthetic cannabinoids 58–60
  synthetic cathinones 98–99
Hofmann reagent 67, 80
HU-210 48–49, 56, 58
HU-211 56, 58

imidazolobenzodiazepines 182, 184
infrared spectroscopy
  cocaine 178
  ketamine detection 207
  lysergic acid diethylamide 71–72
International Drug Control 3
ion mobility spectrometry (IMS),
  fentanyl-type drugs 198

JWH-018 48–50, 58, 59
JWH-022 58, 59

JWH-30 50
JWH-081 56, 57
JWH-175 49, 50
JWH-176 50
JWH-250 50–51
JWH-251 57
JWH-398 57

ketamine
  analogues 208
  gas chromatographic methods
    206–207
  presumptive tests 204
  spectroscopic methods 206–207
  structure 202
  thin layer chromatography
    205–206
ketazolam 183, 189
ketoxime 131
khat analysis 88
  active compounds dose 88
  botanical and microscopic
    analysis 93
  cathinones (*see* cathinones)

LC–MS-based methods, cathinones
  100
Leuckart synthesis, amphetamine
  128
lidocaine 122, 172
Liebermann reagent 172
liquid drug sample 23–24
loprazolam 184
lysergic acid diethylamide (LSD) 19,
  54, 62
  extraction 67
  gas chromatography–mass
    spectrometry 69–70
  gelatin blocks 63
  high-performance liquid
    chromatography 68–69
  LAMPA 65
  paper dose forms 64–65
  physical description 66
  presumptive tests 67
  sample preparation 65
  sampling of dose units 66

spectroscopic methods 70–72
structure 63, 65
thin layer chromatography 67–68

Mandelin reagent 172
Marquis test
 amphetamine-type drugs 132–133
 benzodiazepine 187
 heroin 109
 mescaline analysis 84
 N,N-dimethyltryptamine 80
 piperazines 160
 psilocybin analysis 75
MDAct. See Misuse of Drugs Act
MDMA. See 3,4-methylene-
  dioxymethamphetamine
Mecke test, N,N-dimethyltryptamine 80
mephedrone 89–93, 95, 102
mescaline 63
 cactus material sampling 83
 extraction 83–84
 gas chromatography 85
 packaging and storage 83
 peyote cactus 82
 presumptive tests 84
 thin layer chromatography 84–85
metal ions, heroin 120
methamphetamine 90, 126–128, 132, 137
 diode array spectra 136
 GC–MS analysis 142
methaqualone 29, 105, 109, 112
methoxetamine (MXE) 203
methyl-one 89
methylphenylaziridines 130
methyl tert-butyl ether (MTBE) 71, 162
microcrystal tests
 cocaine 172–173
 phencyclidine and ketamine 205
 piperazines 160
 synthetic cathinones 95–96
Misuse of Drugs Act 1, 3–4, 149, 186, 194, 204
mushroom-based drugs 75

N^6-allyl-6-norlysergic acid
 diethylamide (AL-LAD) 72–73
naphthoylindoles 49, 51
naphthoylpyrroles 51
naphthylmethylindenes 50
naphthylmethylindoles 49
Narcotic Drugs and Psychotropic
 Substances Act 8
N-(β-phenylisopropyl)-benzaldimine 130
N-(β-phenylisopropyl)
 benzylmethylketimine 130
N-ethylhexedrone 89, 92
N-formylamphetamine 128–129
nitrazepam 185, 190
N-methylpropylamide of lysergic
 acid (LAMPA) 65, 68, 71
N,N-di(β-phenylisopropyl)amine 129, 145
N,N-di(β-phenylisopropyl)
 formamide 129
N,N-dimethyltryptamine (DMT) 63
 entheogen 79
 gas chromatography–mass
  spectrometry 80–81
 high-performance liquid
  chromatography 81–82
 NMR analysis 82
 presumptive tests 79–80
 thin layer chromatography 80
N,O-bis(trimethylsilyl)acetamide 43, 114, 115
N,O-bis(trimethylsilyl)
 trifluoroacetamide/
 trimethylchlorosilane (BSTFA/
 TMCS) 44, 56, 76, 77, 176
norpseudoephedrine 93, 94
N-trimethylsilyl-N-methyl
 trifluoroacetamide (MSTFA) 44, 70
nuclear magnetic resonance (NMR)
 spectroscopy
 barbiturates 154
 heroin 116–118
 lysergic acid diethylamide (LSD) 72

Subject Index

N,N-dimethyltryptamine 82
synthetic cannabinoids 60
synthetic cathinones 101
Nyaope 2, 13, 15, 23, 40, 105, 114, 116

ocfentanyl 198, 199
octacosane, chromatogram 116
opiate drugs
    chloroform 114
    heroin samples (see heroin samples)
    legislation 106
oxazepam 190
oxazolam 183
oxazolobenzodiazepines 181, 183

papaverine 104, 107, 111, 113, 116
Papaver somniferum 103, 119
pentobarbital 149, 150
phencyclidine (PCP)
    forms 202
    gas chromatographic methods 206–207
    presumptive tests 204
    spectroscopic methods 206–207
    structure 202
    thin layer chromatography 205–206
phenobarbital 149, 150
phenylacetylindoles 51
photo-oxidation, cannabis 33
piperazines 13, 15, 155
    complex combinations 155, 157
    dissolution 160, 162
    electrochemical methods 166
    gas chromatography–infrared detection 165–166
    gas chromatography–mass spectrometry 163–165
    high-performance liquid chromatography 163
    physical description 159
    presumptive tests 160–161
    structure 156
    synthesis 158

tablet dose forms 157
thin layer chromatography 162
powder drug sample 22–23
procaine 169, 172
psilocin 63, 64, 67, 72
    gas chromatography–mass spectrometry 76–77
    high-performance liquid chromatography 77
    sampling 74
psilocybin analysis 63
    chemiluminescence detection 78
    DNA analysis 78–79
    edible preparations 73–74
    gas chromatography–mass spectrometry 76–77
    hallucinogenic response 72–73
    high-performance liquid chromatography 77–78
    packaging and handling 74
    physical examination 74
    thin layer chromatography 76
    whole dried fungal preparations 74
psychoactive substance 4–5
Psychoactive Substances Act 2016 5, 6
pyrazolam 182, 185

Raman spectroscopy, ketamine detection 207
resinous cannabis material 32–34
    chemical comparison 33–34
    photo-oxidation 33
    physical markings 33
    soap bars 32–33
resinous materials 18–19

secbutabarbital 149
secobarbital 149
Simon's reagent 133
spectroscopic methods
    ketamine 206–207
    phencyclidine and ketamine 206–207
stable isotope analysis

cannabis and products 44–46
cocaine 179–180
heroin 118–120
statistical sampling 21–22
sufentanil 193–195, 201
synthetic cannabinoid receptor
  agonists (SCRAs) 48
synthetic cannabinoids
  adamantoylindenes 51
  benzoylindoles 51
  cyclohexylphenols 48, 49
  gas chromatographic methods
    56–58
  HPLC methods 58–60
  naphthoylindoles 49
  naphthoylpyrroles 51
  naphthylmethylindenes 50
  NMR analysis 60
  packaging 52
  phenylacetylindoles 51
  physical description 53
  plant substrates 51
  presumptive tests 53–54
  tetramethylcyclopropylindoles 51
  thin layer chromatography 54–56
synthetic cathinones
  adverse effect 87
  forms and dosage 92–93
  Fourier transform infrared
    spectroscopy 101
  GC–MS analysis 96–98
  HPLC system 98–100
  LC–MS-based methods 100
  microcrystal tests 95–96
  NMR spectroscopy 101
  presumptive (colour) tests 95
  purity 87
  sampling and physical
    description 94–95
  thin layer chromatography 96
  X-ray crystallography 101

tabletted dose forms 24–25
temazepam 105, 186, 190
tetracosane, chromatogram 116

tetramethylcyclopropylindoles 51
thienodiazepines 182, 184
thienotriazolodiazepines 182, 183
thin layer chromatography (TLC)
  amphetamine-type drugs 133–135
  barbiturates 151
  benzodiazepines 188–189
  cannabis and products 38–40
  cocaine 174–175
  heroin 109–111
  ketamine 205–206
  lysergic acid diethylamide 67–68
  mescaline analysis 84–85
  N,N-dimethyltryptamine (DMT)
    80
  phencyclidine 205–206
  piperazines 162
  psilocybin analysis 76
  synthetic cannabinoids 54–56
  synthetic cathinones 96
total ion chromatograms (TICs)
  191
trace samples, drug recovery 13–16
triazolobenzodiazepines 182, 184
trimethylsilyl-lysergic acid
  diethylamide (TMS-LSD) 70

ultraviolet (UV) spectroscopy,
  cocaine 177–178
United Kingdom Misuse of Drugs
  Act 1971 106
United Nations Office on Drugs and
  Crime (UNODC) 23, 66
UR-144 50, 51, 54, 58

van Urk reagent 54, 67, 80

WIN 55,212-2 56, 58

X-ray crystallography, synthetic
  cathinones 101

Zimmermann reagent
  benzodiazepines 187
  synthetic cathinones 95